TOWARDS A SEMIOTIC BIOLOGY
Life is the Action of Signs

TOWARDS A SEMIOTIC BIOLOGY
Life is the Action of Signs

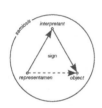  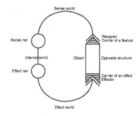

Editors

Claus Emmeche
University of Copenhagen, Denmark

Kalevi Kull
University of Tartu, Estonia

Imperial College Press

Published by

Imperial College Press
57 Shelton Street
Covent Garden
London WC2H 9HE

Distributed by

World Scientific Publishing Co. Pte. Ltd.
5 Toh Tuck Link, Singapore 596224
USA office: 27 Warren Street, Suite 401-402, Hackensack, NJ 07601
UK office: 57 Shelton Street, Covent Garden, London WC2H 9HE

British Library Cataloguing-in-Publication Data
A catalogue record for this book is available from the British Library.

TOWARDS A SEMIOTIC BIOLOGY
Life is the Action of Signs

Copyright © 2011 by Imperial College Press

All rights reserved. This book, or parts thereof, may not be reproduced in any form or by any means, electronic or mechanical, including photocopying, recording or any information storage and retrieval system now known or to be invented, without written permission from the Publisher.

For photocopying of material in this volume, please pay a copying fee through the Copyright Clearance Center, Inc., 222 Rosewood Drive, Danvers, MA 01923, USA. In this case permission to photocopy is not required from the publisher.

ISBN-13 978-1-84816-687-5
ISBN-10 1-84816-687-7

Typeset by Stallion Press
Email: enquiries@stallionpress.com

Printed in Singapore.

Contents

Preface ix

List of Contributors xi

Chapter 1: Why Biosemiotics? An Introduction to
Our View on the Biology of Life Itself 1
Kalevi Kull, Claus Emmeche and Jesper Hoffmeyer

Part I: Biosemiotic Approach: General Principles

Chapter 2: Theses on Biosemiotics: Prolegomena to
a Theoretical Biology 25
*Kalevi Kull, Terrence Deacon, Claus Emmeche,
Jesper Hoffmeyer and Frederik Stjernfelt*

Chapter 3: Biology Is Immature Biosemiotics 43
Jesper Hoffmeyer

Chapter 4: Biosemiotic Research Questions 67
Kalevi Kull, Claus Emmeche and Donald Favareau

Chapter 5: Organism and Body: The Semiotics of
Emergent Levels of Life 91
Claus Emmeche

Chapter 6: Life Is Many, and Sign Is Essentially
Plural: On the Methodology of Biosemiotics 113
Kalevi Kull

Part II: Applications

Chapter 7: The Need for Impression in the
Semiotics of Animal Freedom:
A Zoologist's Attempt to Perceive
the Semiotic Aim of H. Hediger 133
Aleksei Turovski

Chapter 8: The Multitrophic Plant–Herbivore–
Parasitoid–Pathogen System:
A Biosemiotic Perspective 143
Luis Emilio Bruni

Chapter 9: Structure and Semiosis in Biological Mimicry 167
Timo Maran

Chapter 10: Semiosphere Is the Relational Biosphere 179
Kaie Kotov and Kalevi Kull

Chapter 11: Why Do We Need Signs in Biology? 195
Yair Neuman

Part III: Conversations

Chapter 12: Between Physics and Semiotics 213
Howard H. Pattee and Kalevi Kull

Chapter 13: A Roundtable on (Mis)Understanding of Biosemiotics 235

Claus Emmeche, Jesper Hoffmeyer, Kalevi Kull, Anton Markoš, Frederik Stjernfelt, Donald Favareau

Chapter 14: Theories of Signs and Meaning: Views from Copenhagen and Tartu 263

Jesper Hoffmeyer and Kalevi Kull

Acknowledgements	287
Name Index	289
Subject Index	297

Preface

Understanding life will never be an easy task. But it is always a fascinating exercise.

Our path in this search to understand the life processes has led us, as biologists, to a semiotic view. Life processes are not only significant for the organisms they involve. Signification, meaning, interpretation and information are not just concepts used and constructed by humans for describing such processes. We conclude that life processes themselves, by their very nature, are meaning-making, informational processes, that is, sign processes (semioses), and thus can be fruitfully understood within a semiotic perspective.

This book presents a series of essays written collectively or under a mutual influence by biosemioticians who have gathered at seminars and meetings held in Copenhagen and Tartu, as well as in some other corners of the world.

The selection of texts here does not cover the entire scope of views that have been developed in the last couple of decades under the name of biosemiotics or semiotic biology. As in general semiotics, in biosemiotics we can find several theoretical preferences represented by different scholars. The Copenhagen and Tartu groups are close to each other and share many of their theoretical views.

For several of the texts presented here, earlier versions have been published in different places. In these cases they are re-edited, with some sections appearing for the first time. The introductory article is written especially for the current volume. The rest of the book is organised into three parts. Part I focuses on the main conceptual issues. Part II provides examples of more data-driven and applied research. Part III includes discussions presented in dialogic format, providing self-reflections of the field.

The whole of this book seems to us to be a small step forward in a long process, in which different groups of biologists work to develop a more adequate paradigm for a scientific grasp of life. To this process belong the meetings of the *Theoretical Biology Club* of the 1930s in Cambridge, the *Towards a Theoretical Biology* symposia at Lake Como in 1968–1972, and the *Gatherings in Biosemiotics* in Copenhagen, Tartu, and several other places, since 2001. The latter could not happen without the earlier efforts of Thomas Sebeok, Thure von Uexküll, and Jesper Hoffmeyer. There have been, of course, similar events before, and we look forward to the ones in the future.

Understanding life is a life's delight. Helpful, too.

Claus Emmeche
Kalevi Kull
Spring 2010, Copenhagen and Tartu

Contributors

Luis Emilio Bruni

Department of Media Technology and Engineering Science,
Faculty of Engineering, Science and Medicine, Aalborg University,
Lautrupvang 15, 2750 Ballerup, Denmark
e-mail: leb@imi.aau.dk

Terrence Deacon

Department of Anthropology, University of California, USA
e-mail: deacon@berkeley.edu

Claus Emmeche

Center for the Philosophy of Nature and Science Studies,
Niels Bohr Institute, University of Copenhagen,
Blegdamsvej 17, 2100 Copenhagen, Denmark
e-mail: emmeche@nbi.dk

Donald Favareau

University Scholars Programme,
National University of Singapore, Singapore
e-mail: favareau@biosemiotics.org

Jesper Hoffmeyer

Department of Molecular Biology,
University of Copenhagen, Denmark
e-mail: jhoffmeyer@me.com

Kaie Kotov
Department of Semiotics, University of Tartu,
Tiigi 78, 50410 Tartu, Estonia
e-mail: kaie.kotov@gmail.com

Kalevi Kull
Department of Semiotics, University of Tartu,
Tiigi 78, 50410 Tartu, Estonia
e-mail: kalevi.kull@ut.ee

Timo Maran
Department of Semiotics, University of Tartu,
Tiigi 78, 50410 Tartu, Estonia
e-mail: timo.maran@ut.ee

Anton Markoš
Department of Philosophy and History of Science,
Faculty of Sciences, Vinicna 7, Charles University,
CZ 128 00 Praha 2, Czech Republic
e-mail: markos@natur.cuni.cz

Yair Neuman
Ben Gurion University, Israel
e-mail: yneuman@bgu.ac.il

Howard H. Pattee
1611 Cold Spring Road, Williamstown, MA 01267, USA
e-mail: hpattee@roadrunner.com

Frederik Stjernfelt
Center for Semiotics, Aarhus University, Denmark
e-mail: semfelt@hum.au.dk

Aleksei Turovski
Tallinn Zoo, Tallinn, Estonia
e-mail: zoosemu@hotmail.com

Chapter 1
Why Biosemiotics? An Introduction to Our View on the Biology of Life Itself

Kalevi Kull, Claus Emmeche and Jesper Hoffmeyer

Summary. This chapter introduces the very idea of a semiotic biology. Here, one needs to know what semiosis, the action of sign, is, and why semiotics, the study of signs processes, provides a strong conceptual toolbox to approach a more complete theoretical biology. Sebeok's thesis that living systems are constituted as sign systems is a key point of departure for the emergent science of biosemiotics. The history of semiotics in the 20th century has been influenced deeply by structuralism in linguistics, and this "semiology" is related to a similar structuralist movement in theoretical biology. This, together with a variety of approaches that emphasize the properties of relation, signification, wholeness and contextuality, can be seen as forerunners to a more fully developed semiotic biology, that sees living creatures not just as passively subjected to universal laws of nature, but also as active systems of sign production, sign mediation and sign interpretation, that harness the physical laws in order to live and sometimes to make a more complex living. Such a biology is not in conflict with present-day biological research, but it promises to be a good guide towards a theoretical biology that does not turn the emergence of life — and with it, meaning and intentionality — into an incomprehensible mystery. Thus, biosemiotics would provide the basis for linking general biology with general linguistics.

To be alive is to be semiotically active.
Thomas A. Sebeok

Sebeok's thesis

"Semiosis is what distinguishes all that is animate from lifeless" (Sebeok 1988: 1089). We have named this statement *Sebeok's thesis*, since Thomas A. Sebeok, an architect of biosemiotics, made it widely known by stressing it repeatedly (Kull, Emmeche, Favareau 2008: 43). However, the thesis has been seen earlier, though worded differently. For instance, Friedrich Rothschild wrote that "living systems are constituted from their very beginning as sign systems" (Rothschild 1968: 174). If Sebeok's thesis is true, then it implies that a theory of semiosis, contemporarily called *semiotics*, would include the entire sphere of *biology* (to be precise — that part of biology which deals with *living* systems).

Semiosis is the sign process — the fundamental process that carries meaning and in which meaning is created. It is the process — not at all simple — that mediates purpose and causality, living and dead aspects of nature, and makes it possible to see how to overcome a crude dualism of mind and matter, as well as how the dynamics of the actions of signs provides a better approach to living systems than our dichotomies of mental versus physical properties. In semiotic investigations of this process, a series of models to describe its peculiarities has been worked out (for an early review, see Krampen 1997).

An important and necessary assumption that makes space for semiotics in biology, i.e., outside of the study of human cultural sphere, is that of the existence of meaningful communication in many other species besides *Homo sapiens*. Animal communication has been studied from this point of view and described in the framework of zoosemiotics since 1963. The organisms' interpretation of their world — *umwelt*, as studied by Jakob von Uexküll as early as in the 1920s — is also a semiotic process.

However, the idea of applying the theories and concepts of semiotics to "natural" phenomena in non-human life has not easily been accepted by all semiotic researchers. For Ferdinand de Saussure, whose *Course in General Linguistics* was published in 1916, semiology (the science of signs as he called it) was a field that would cover a much wider range of sign systems than human language. Half a century later, Roland Barthes, a very influencial semiologist from whom many directions of research were initiated since the start of the institutionalization of semiotics in the 1960s, has pointed out that the sphere of sign phenomena that seemed to be non-linguistic have all been later understood as based on the human language capacity. Therefore, Barthes (1997 [1967]: 11) says: "In fact, we must now face the possibility of inverting Saussure's declaration: linguistics is not a part of the general science of signs, even a privileged part, it is semiology which is a part of linguistics: to be precise, it is that part covering the great signifying unities of discourse." This has narrowed the scope of a science of signs in many schools for quite a long time.

Simultaneously with Barthes, Thomas Sebeok moved semiotic theory in another direction, including not only human non-verbal signs (which may, indeed, be based on or at least influenced by the language capacity), but also animal communication. Leaving the term "language" specifically for the human symbolic sign system, Sebeok and his colleagues thus made general semiotics a truly and fundamentally supralinguistic field.

In the last half century semiotics has developed remarkably. Biosemiotics has become a necessary part of contemporary sign systems studies. Among the recent advancements of biosemiotics we can list the operational distinction between semiosic and non-semiosic, codes and stereochemical relationship,[1] communication

---

[1] The threshold between semiosic and non-semiosic as between code-relations and stereochemical relationships (and accordingly the usage of semiotic terminology — like "translation", "interpretation", "coding", etc. — either in a direct or in a metaphorical sense) was formulated by Umberto Eco (1986: 183; 1988). See discussion on this in Kull (2003: 596–597).

and non-communicational interaction — all this has been an important achievement for the clear understanding of the scope of the semiotic sphere. The efforts of biosemioticians have advanced not only the conceptual apparatus of general semiotics, but also biosemiotics itself — the theoretical apparatus for the description of biological phenomena in all levels of organization of life. The last decade or two have institutionalized biosemiotics as an academic speciality (with a society, a journal, regular scientific meetings, etc.) with highly transdisciplinary ambitions.

Thus, biosemiotics can no longer be seen as just a marginal school of thought at some universities, but as a new biology about to find its proper theoretical ground (cf. Hoffmeyer 2009). If semiotics, because of the wide scope of a scientific study of sign processes, can provide a fundamental basis for all sciences dealing with life, then biosemiotics will not only have a major theoretical impact upon ethology and behavioral ecology, fields traditionally devoted to the investigation of animal communication, but may, in the long run, restructure much of the institutional map of science and scholarship.

The development of semiotic thinking in the 20th century has been described as not only a scientific matter, but also a cultural phenomenon, where change encompasses deep aspects of worldview. In biology, the introduction of a semiotic approach has had its predecessors, among which we would like to point here to the structuralist trend. Linguistic structuralism has prepared the rise of semiotics, as biological structuralism has prepared biosemiotics. And as it seems, these two types of structuralism have had some mutual influence — still not yet studied by the historians of ideas.

Structuralism as a step towards semiotics

One of the long-term approaches that has appeared very much in a connection with sign systems studies is structuralism. Structuralism is an approach that is an aspect of semiology, or Saussurean semiotics.[2] Structuralism has been extremely important in the development of semiotic thinking in the 20th century, gaining a wide popularity in the 1960s and 1970s in many fields of the humanities. However, it is only one part, mainly a non-dynamic part of semiotics, which was largely overcome in the 1980s when the wider usage of the concept of semiosis (as a general phenomenon of sign action and interpretation at various levels of signification) supplemented, and to some extent, gave the concept of code — a core concept of semiotic theory and thinking — a more dynamic content.

The term "structuralism" was introduced by Roman Jakobson in 1929:

> Were we to comprise the leading idea of present-day science in its most various manifestations, we could hardly find a more appropriate designation than *structuralism*. Any set of phenomena examined by contemporary science is treated not as a mechanical agglomeration but as a structural whole, and the basic task is to reveal the inner, whether static or developmental, laws of this system. What appears to be the focus of scientific preoccupations is no longer the outer stimulus, but the internal premises of the development; now the mechanical conception of processes yields to the question of their functions. (Jakobson 1971 [1929]: 711)

Structuralism puts a strong emphasis upon the importance of synchronic rather than diachronic factors that make structures. It is also based on binary oppositions, and emphasises the importance of codes and meaning (see, e.g., Schogt 1994).

Structuralism in linguistics, seeing everyday use of language as guided by some deeper principles of a relational system and envisioning the possibility of extending the notion of structures of significance beyond language to other cultural phenomena, has

---

[2]See, e.g., Bouissac (2010).

been an important step in the formation of a semiotic approach that looks for relations of signification even beyond the human domain of social structures. One of the central tenets of structuralism concerns the relational nature of a system like the system of language. The system consists not simply of an ontology or a vocabulary of elementary units, particles, or entities; what is important is the relations of similarity and difference between the entities, and the fact that their place within the system defines their role, significance, or function. Individual entities may be substituted, or transformed, into other entities like words in different national languages, but the code defining the rules by which the entities relate to each other may still be the same. Particular material properties of such entities, like the wide range of ways to pronounce a word like "dog" in English, may not be of central interest. But imagine a continuous scale of 50 slightly variant sounds from some version of "dog" to some version of "vogue"; these sounds can be tested for when the meaning shifts from "dog" to "vogue". What defines a phoneme is not its physics or its sound as such, but its relation of difference to other, similar phonemes.

Thus, one of the early findings of structuralism was that the system of phonemes is not a consequence of physical laws, but it forms an interrelated system of differences, based on communicative recognition between speakers. Similarly, in biology, the system of biological species can be seen as relational in a similar way, as based on mutual recognition in the process of biparental reproduction. This fundamental analogy of phonological and speciational mechanisms has been noted by some biologists and linguists (Kull 1992; Goodenough 1992) and deserves attention in semiotics.

We can even see the main principles of structuralism present in classical biological taxonomical work (a representative of which was, for instance, Ferdinand de Saussure's father, entomologist Henri Saussure), which did not use an evolutionary explanation but instead developed a structural understanding of a natural system of taxa. Similarly, in some theories in ecology dealing with succession, communities, and niches, we see a focus upon the ecological roles played by, e.g., primary producers, primary consumers, secondary consumers, etc., where the structure of a functional system of relations is more important than the particular physics of its elements.

In biology, the structuralist perspective has been applied especially in the studies of organic form. Remarkable works directly identifying themselves as structuralist were developed by René Thom (1972) and the Osaka Group (Goodwin *et al.* 1989; Webster 1989; Sermonti, Sibatani 1998; see also Depew, Weber 1997: 418–419, Markoš 2002: 81–88). These studies share much with the biosemiotics of the 1990s and 2000s.

First, structuralism is anti-reductionist. René Thom (1972: 71) wrote:

> From a qualitative point of view, [structuralist methods] represent an analogous position to the famous *hypotheses non fingo* of Newton. In the structuralist viewpoint, one does not try to explain a morphology by reduction to elements borrowed from another theory — supposedly more elementary or fundamental — as one might try to explain biology by physics and/or chemistry, or sociology by psychology or biology; one only tries to improve the description of the empirical morphology by exhibiting its regularities, its hidden symmetries, by showing its internal unity through a formal mathematical model which can be generated automatically. In that respect "structuralism" is a modest theory, as its only purpose is to improve description.

Second, structuralism supports a non-Darwinian evolutionism.[3] This theme has been well developed by the Osaka group (Goodwin *et al.* 1989). For instance, Gerry Webster (1989: 2–3)

---

[3] See also Dwyer (1984).

wrote:

> From [...] Lévi-Strauss' anthropology we can extract a series of conceptual oppositions which are relevant to the problem of biological form:
> (1) internal constraints *versus* external demands;
> (2) reproduction *versus* inheritance;
> (3) law-governed transformation *versus* accidental or random variation;
> (4) a "rational" system of transformations (a structure) *versus* a purely "empirical" (temporal or spatial) order.
>
> Now, in relation to biology, if the first member of each pair of concepts is structuralist, then the second member is certainly Darwinist in the sense in which that term was used in the late nineteenth century, and to a considerable extent is used today, to refer to the theory of descent plus the theory of natural selection. This opposition of structuralism and Darwinism is not simply a debating device. In the first place, it is of historical significance for a number of earlier writers ([William] Bateson, Driesch, D'Arcy Thompson) who employed structuralist concepts did so in the course of formulating critical evaluations of Darwinism.

This long quotation illustrates, among other things, the relatedness of linguistic (Saussurean) and biological structuralism. The same biological structuralism has been a central topic in Stephen Jay Gould's *magnum opus* (Gould 2002). He uses intermittently the terms structuralism and formalism. He claims: "*Formalism* and functionalism represent poles of a timeless dichotomy" (Gould 2002: 312). The functionalists were Georges Cuvier and Charles Darwin, formalists Etienne Geoffroy Saint-Hilaire and Richard Owen. Gould's position was clearly critical of the functionalists.

It is worth noting that several structuralist biologists have directly stated the relatedness of their view to linguistic structuralism — Thom (1972), Webster (1989), Jiménez-Montano (1992), Barlow and Lück (2007). It is also very interesting that in the 1930s, Roman Jakobson, in the process of becoming an architect of later structuralism, used Karl Ernst von Baer's and Lev Berg's non-Darwinian biological work as major sources for working out his approach to linguistics during his years in Prague (Sériot 1995; 2001; 2003).

Interestingly, we can find something similar in Uexküll. Ernst Cassirer (1950: 200) observed: "While Driesch in his conception of entelechy wanted to demonstrate a specific autonomy of function, Uexküll started from the autonomy of *form*." And as stated by Jakob von Uexküll (1930: 9), in Cassirer's (1950: 200) translation:

> Structure is not a material thing: it is the unity of immaterial relationships among the parts of an animal body. Just as plane geometry is the science not of the material triangles drawn on a blackboard with chalk but of the immaterial relationships between the three angles and three sides of the closed figure [. . .] so biology treats of the immaterial relationships of material parts united in a body so as to reconstitute the structure in imagination.[4]

In their later book, Webster and Goodwin (1996) use the terms "generative biology" and "relational biology" instead of "structuralist". However, they connect the principle of "generating relation" with Ernst Cassirer's philosophical structuralism (Webster, Goodwin 1996: 104–109). Further, attention is turned towards complexity and its models (Solé, Goodwin 2000). Without a direct reference, they as well as other theoretical biologists probably know that "relational biology" is what made a most significant turn for Nicolas Rashevsky

---

[4] A comment as to the term "immaterial" would be useful here. Relations are immaterial, and yet they are concerned with material entities (*ens reale* as the scholastic called it). So immaterial, yes, in a way, but with material consequences.

and Robert Rosen:

> But complexity, though I suggest it is the habitat of life, is not itself life. Something else is needed to characterize what is alive from what is complex. Rashevsky provided this too, in his idea that biology was relational, and that relational meant (as we stated it) throwing away the physics and keeping the organization. A rough analog would be: throwing away the polypeptide and keeping the active sites. Organization in its turn inherently involves functions and their interrelations; the abandonment of fractionability, however, means that there is no kind of 1 to 1 relationship betweeen such relational, functional organizations and the structures which realize them. These are the basic differences between organisms and mechanisms or machines. (Rosen 1991: 280)

Rosen's approach is a path towards biosemiotics.

A structuralist view, importantly, assumes a certain independent existence of structural laws from the material sphere of phenomena or objects. This also has several manifestations within semiotic approaches to biology. Where they have not been properly embedded into a semiotic theory and there has been an attempt to develop them separately and independently, semiological or structuralist approaches have led to a one-sidedness (a *pars pro toto* fallacy, see Deely 2009) that a biosemiotic approach avoids. We observe several deviating trends:

(a) geometrization of organic form (as well-developed by D'Arcy Thompson in 1942 — which is one of the landmarks of 20th century morphology) has been a tendency in several structuralist treatises of morphogenetics, including those with semiotic emphasis (e.g., Zarenkov 2007; cf. Gutmann 2004);
(b) equation of organic and inorganic regularities; this often means a view that there are no universal laws, all laws are in principal of the same type, and the methodology of exact sciences

is appropriate to use for life sciences *and* humanities, i.e., for semiotics (about this see also Markoš 2002: 84);

(c) reduction of sign relations to the existence of codes (Barbieri 2001); then semiosis tends to be a result of codes and not vice versa (where codes could arise as habits from more dynamic sign relations);

(d) preference for binary oppositions instead of triadic (or plural) relations (Florkin 1974); this may correspond to a view on the primacy of opposition instead of fundamental plurality of meaning relations;

(e) limiting the semiotic approach to the level of scientific description only, and not to the level of reality of the living system itself; in this case a semiotic biology would mean exclusively the semiotics of doing biology, and not the semiosic aspects of living systems.

In some cases, structuralism has noticed its biological basis. Structuralism "can only appear in the mind because its model is already present in the body" (Lévi-Strauss 1981: 692; see also Dosse 1997a: 389). "Lévi-Strauss's ambition during the fifties to belong to the natural sciences became a program during structuralism's second phase; biology replaced linguistics" (Dosse 1997b: 397).

Thus, in a way we repeat here a claim of Ernst Cassirer from his article "Structuralism in modern linguistics" (1945), where "he made a crucial analogy between morphology in biology and structuralism in linguistics" (Smith 2004: 4; see also Krois 2004).

Several additional historical aspects of the development from structuralism towards semiotics have been described by Nöth (1990: 295ff).[5]

---

[5]Hawkes (2005 [1977]: 101) remarked that, by and large, the boundaries of semiotics "are coterminous with those of structuralism: the interests of the two spheres are not fundamentally separate and, in the long run, both ought properly to be included within the province of a third, embracing discipline called, simply, *communication*." This is what has indeed happened

More steps towards semiotics

We devoted the remarks above to structuralism because a large part of structuralism has been strongly connected to semiology, whereas semiology can be seen as a certain part of semiotics. From a semiotic point of view, semiology is based on linguistic, dyadic and quite static models of sign relation, whereas semiotics stands on more general triadic and dynamic models. Therefore, identification of semiology with semiotics would be wrong. Semiological models represent the restricted special cases of semiotic models.

From the point of view of contemporary semiotics, not only in structuralism, but also in functionalism, some ideas can be interpreted as a preparation for semiotics. For instance, one may recall Adolf Meyer-Abich's claim about Uexküll; according to him, Uexküll's classification of functional circles was the first attempt in the history of biology to describe the functional archetypes (Meyer-Abich 1963). Many models of intentionality, learning, adaptation, and habit-taking belong obviously to problems of the functionalist sphere, while describing the important aspects of sign processes.

Semiotics as the science of sign processes is quite extensive already, due to its object. Even in a narrow sense as a study of symbols (which is only one type of signs, specific to humans), the traditions of semantics, hermeneutics, logic, and rhetoric have all contributed to semiotics (Nöth 1990: 11). All these have representatives on the way towards biosemiotics.

(a) *Biohermeneutics* has been developed by Sergey Chebanov (1993; 1999) and Anton Markoš (2000).[6] Hermeneutics is an

---

in many American universities — however, with a considerable loss of the depth of semiotic theory. In this case "semiotics" will be used in a narrow, Barthes-like sense (Hawkes 2005: 101 fn3). Our usage, which follows the tradition of the International Association for Semiotic Studies, is a broad one, according to which structuralism is only a part and a method of semiotics.

[6] The initial subtitle of Markoš book — the translation from Czech — was "The hermeneutics of the living", but as often happens with publishing houses that depend on the market, the author had to change it.

interpretation of interpretation. Its core understanding sees the closure of this procedure, i.e., the endlessness of interpretation.[7]

Markoš (2002: 35) states: "hermeneutics is the very act of *acquiring* knowledge, understanding, becoming acquainted with a given thing, that is, *creating* meaning." He quotes Hans-Georg Gadamer (1996 [1960]: 225–226; see Markoš 2002: 49): "[For Wilhelm Dilthey,] significance is not a logical concept, but is to be understood as an expression of life. Life itself, flowing temporality, is ordered toward the formation of enduring units of significance. Life interprets itself. Life itself has a hermeneutic structure. Thus life constitutes the real ground of the human sciences." Markoš applies hermeneutics in biology. This is quite close to the idea that life is a self-reading text, and that an aim of biosemiotics is to know what other organisms know (Kull 2009).

(b) Ruth Millikan (1989), Walter Fontana (1994), and Marcello Barbieri (2001) have tried to build *biosemantics* — each of them on a quite different theoretical basis.

(c) Stephen Pain has made attempts to develop *biorhetorics* (Pain 2002; 2009; see also Kull 2001).

(d) Also, it makes sense to speak about logic at the level of animal, and possibly also vegetative life — on *bio-logic* — asking, for instance, whether it is possible to have a certain type of logical contradiction based on semiosic processes already at these levels of life. Such an approach would correspond well to Charles S. Peirce's interpretation of logic. Cooper (2001) has written about the biological roots of logic, although approaching the matter by a different route.

(e) The name *biolinguistics* has been used for studies on the biological basis of human language, or the study of relations between physiology and speech. Despite many parallels between biolinguistics and biosemiotics (see Augustyn 2009), these traditions are quite different. Biolinguistics tries to apply

---

[7] On the relationship between semiotics and hermeneutics, see, for example, Niklas 1991.

the existing (physicalistic) biology in linguistics. Biosemiotics finds it necessary to put biology on a semiotic basis before integration of the theories of biology and linguistics.

The biosemiotic approach

Already in the 1960s, when semiotics began to institutionalize and textbooks and introductions started to be published for the introductory and general courses of semiotics, the biological sphere as represented by animal communication studies (zoosemiotics) and biosemiotics (which was coined as a term only in 1962) became a part of semiotics as a field (as, for instance, reflected in the first introductory books to semiotics by Bense 1967, Vetrov 1968, Mounin 1970, Stepanov 1971). However, until the early 1990s, biosemiotic work was quite scarce. Jesper Hoffmeyer (2008a: 3) observes: "This investigation into the semiotic nature of living systems has taken a long time to emerge, since it poses a challenge to many of the prevailing ontological assumptions of both the natural and the human sciences."

From the early 1990s on, biosemiotics started to produce books under its own name, with Thomas Sebeok and Jesper Hoffmeyer as the leading scholars (Sebeok, Umiker-Sebeok 1992; Sebeok 1990; Hoffmeyer 1996; 2008a; 2008b) and with many other in support (Deacon 1997; Merrell 1996; Witzany 1993; 2007; Cimatti 1998; Barbieri 2001; 2007; Markoš 2002; Emmeche *et al.* 2002; Weber 2003; Andrade Pérez 2003; Schult 2004; Wheeler 2006; Stjernfelt 2007; Martinelli 2007; Maran 2008; Brier 2008; El-Hani *et al.* 2009). Several journals have published special issues on biosemiotics (*Semiotica* 89(4), 1992; 120(3/4), 1998; 127(1/4), 1999; 134(1/4), 2001; *Sign Systems Studies* 30(1), 2002; 32(1/2), 2004; 37(3/4), 2009; *Cybernetics and Human Knowing* 5(1), 1998; 7(1), 2000; 10(1), 2003; *Zeitschrift für Semiotik* 8(3), 1986; 15(1/2), 1993; 18(1), 1996; *European Journal of Semiotics* 9(2), 1997; *American Journal of Semiotics* 24(1/3), 2008; *Cognitive Semiotics* 4, 2009). Specialized journals (*Journal of Biosemiotics*, 2005; *Biosemiotics*,

2008–), and a book series, *Biosemiotics*, with Springer, have been established in recent years.

Semiotics, of course, has a long and rich history (Posner *et al.* 1997; Deely 2001), and so has biosemiotics (see Favareau 2007; 2010; Kull 1999). However, semiotics came onto the scene with its full meaning, sense, and power only once we started to investigate how life actually works, only with the impetus to understand the building blocks of life. This is particularly true since the "symbol grounding problem" (Harnad 1990; Cangelosi *et al.* 2002) can be said to be solved at least with some models of semiosis.[8] Obviously, these results are influenced by the extensive knowledge collected in molecular and cellular biology and neurobiology about the composition of living systems.

Thus, semiotics is not only a matter of description, it is also primarily a matter of mechanisms — of semiosic mechanisms. The semiotic approach, being broader than the structuralist, would not see the binary oppositions as fundamental, binarism rather belonging to the linguistic sphere. Instead of codes, the general concept will be semiosis. The intentional aspect of sign processes, which parallels the teleological–teleonomical aspect in life processes, has already been considered in structuralism, but starts to play a more remarkable role in semiotics. Accordingly, biosemiotics is *biology as a sign systems study*, or, in other words, a *study of semiosis in living nature*. "Biosemiotics is the name of an interdisciplinary scientific project that is based on the recognition that life is fundamentally grounded in semiotic processes" (Hoffmeyer 2008a: 3).

Any biological system is a web of linked recognition processes. The cell, often considered a basic unit of life, has to rely upon energy

---

[8]Furthermore, the key problem — that of grounding the intentional content of symbols in their existence as parts of a so-called "physical symbol system" — dissolves from the point of view of another, more developed and genuinely semiotic, notion of symbol as a special sign mode, like the one we find in Peirce.

from its surroundings to keep its metabolism going, and useful energy, substances containing useful energy, has to be distinguished from detrimental substances or influences which are too energetic and may destroy the internal workings of the cell. Conceptually, the fundamental border between cell and non-cell, or self and non-self, is a border that physically is (as if) realized by the membrane of the cell, but that is functionally dependent upon recognition processes and thus structured as a simple code or system of categorization, where the cell recognizes some substrates as "relevant", some as dangerous and should be avoided, and some as simply neutral. This borderline between cell and surroundings is the original informational "difference that makes a difference to the organism", and an elaborate system of macromolecules work as semiotic mechanisms supporting this boundary work, via continuous effort, to maintain the difference between inside and outside. Since creation of organic functions, sign relations begin.

Sign relations are those that connect living systems at any level. These are responsible for communication, speciation, differentiation, and qualitative diversity. We can imagine a semiotic theory of evolution.

Among the primary problems for science in any field is an explanation of diversity and stability. So in semiotics — mechanisms of stability in living systems are semiosic. "Semiosis is, in fact, *the* instrument which assures the maintenance of the steady state of any living entity, whether in Liliputian microspace, dealt with by molecular geneticists and virologists; the Gulliver-sized world of our daily existence; or [...] the biosphere viewed as a Brobdingnagian macrostructure that subsists upon a splendid blue marble" (Sebeok 1988: 1085).

References

Andrade Pérez, Luis Eugenio 2003. *Los Demonios de Darwin: Semiótica y Termodinámica de la Evolución Biológica*. 2nd ed. Bogotá: Universidad Nacional de Colombia.

Augustyn, Prisca 2009. Uexküll, Peirce, and other affinities between biosemiotics and biolinguistics. *Biosemiotics* 2(1): 1–17.
Barbieri, Marcello 2001. *The Organic Codes: The Birth of Semantic Biology*. Ancona: Pequod.
Barbieri, Marcello (ed.) 2007. *Introduction to Biosemiotics. The New Biological Synthesis*. Dordrecht: Springer.
Barlow, Peter W.; Lück, Jacqueline 2007. Structuralism and semiosis: Highways for the symbolic representation of morphogenetic events in plants. In: Witzany, Günther (ed.), *Biosemiotics in Transdisciplinary Contexts*. Vilnius: Umweb, 37–50.
Barthes, Roland 1997 [1967]. *Elements of Semiology*. New York: Hill and Wang.
Bense, Max 1967. *Semiotik: Allgemeine Theorie der Zeichen*. Baden-Baden: Agis.
Bouissac, Paul 2010. *Saussure: A Guide for the Perplexed*. London: Continuum.
Brier, Soren 2008. *Cybersemiotics: Why Information is not Enough*. Toronto: University of Toronto Press.
Cangelosi, Angelo; Greco, Alberto; Harnad, Stevan 2002. Symbol grounding and the symbolic theft hypothesis. In: Cangelosi, Angelo; Parisi, Domenico (eds.), *Simulating the Evolution of Language*. London: Springer, 191–210.
Cassirer, Ernst 1945. Structuralism in modern linguistics. *Word. Journal of the Linguistic Circle of New York* 1(2): 99–120.
Cassirer, Ernst 1950. *The Problem of Knowledge: Philosophy, Science, and History since Hegel*. New Haven: Yale University Press.
Chebanov, Sergey V. 1993. Man as participant to natural creation: Enlogue and ideas of hermeneutics in biology. *Rivista di Biologia* 87(1): 39–55.
Chebanov, Sergey V. 1999. Biohermeneutics and hermeneutics of biology. *Semiotica* 127(1/4): 215–226.
Cimatti, Felice 1998. *Mente e linguaggio negli animali: Introduzione alla zoosemiotica cognitiva*. Roma: Carocci editore.
Cooper, William S. 2001. *The Evolution of Reason: Logic as a Branch of Biology*. Cambridge: Cambridge University Press.
Deacon, Terrence 1997. *The Symbolic Species*. London: Penguin.
Deely, John 2001. *Four Ages of Understanding*. Toronto: University of Toronto Press.
Deely, John 2009. Pars pro toto from culture to nature: An overview of semiotics as a postmodern development, with an anticipation of developments to come. *American Journal of Semiotics* 25(1/2): 167–192.
Depew, David J.; Weber, Bruce H. (eds.) 1997 [1995]. *Darwinism Evolving: Systems Dynamics and the Genealogy of Natural Selection*. Cambridge: MIT Press.
Dosse, François 1997a [1991]. *History of Structuralism: vol. 1, The Rising Sign, 1945–1966*. (Glassman, Deborah, trans.) Minneapolis: University of Minnesota Press.
Dosse, François 1997b [1992]. *History of Structuralism: vol. 2, The Sign Sets, 1967–present*. (Glassman, Deborah, trans.) Minneapolis: University of Minnesota Press.

Dwyer, Peter D. 1984. Functionalism and structuralism: Two programs for evolutionary biologists. *The American Naturalist* 124(5): 745–750.
Eco, Umberto 1986. *Semiotics and the Philosophy of Language*. Bloomington: Indiana University Press.
Eco, Umberto 1988. On semiotics and immunology. In: Sercarz, Eli E.; Celada, Franco; Michison, N. Avrion; Tada, Tomio (eds.), *The Semiotics of Cellular Communication in the Immune System*. Berlin: Springer, 3–15.
El-Hani, Charbel Niño; Queiroz, João; Emmeche, Claus 2009. *Genes, Information, and Semiosis*. (Tartu Semiotics Library 8.) Tartu: Tartu University Press.
Emmeche, Claus; Kull, Kalevi; Stjernfelt, Frederik 2002. *Reading Hoffmeyer, Rethinking Biology*. (Tartu Semiotics Library 3.) Tartu: Tartu University Press.
Favareau, Donald 2005. Founding a world biosemiotics institution: The International Society for Biosemiotic Studies. *Sign Systems Studies* 33(2): 481–485.
Favareau, Donald 2007. The evolutionary history of biosemiotics. In: Barbieri, Marcello (ed.), *Introduction to Biosemiotics: The New Biological Synthesis*. Berlin: Springer, 1–67.
Favareau, Donald (ed.) 2010. *Essential Readings in Biosemiotics: Anthology and Commentary*. Berlin: Springer.
Florkin, Marcel 1974. Concepts of molecular biosemiotics and of molecular evolution. *Comprehensive Biochemistry* 29A: 1–124.
Fontana, Walter 1994. Molekulare Semantik: Evolution zwischen Variation und Konstruktion. In: Braitenberg, Valentin; Hosp, Inga (eds.), *Evolution: Entwicklung und Organisation in der Natur*. Reinbek bei Hamburg: Rowohlt, 69–106.
Goodenough, Ward H. 1992. Gradual and quantum changes in the history of Chuukese (Trukese) phonology. *Oceanic Linguistics* 31(1): 93–114.
Goodwin, Brian; Sibatani, Atuhiro; Webster, Gerry (eds.) 1989. *Dynamic Structures in Biology*. Edinburgh: Edinburgh University Press.
Goodwin, Brian 1992 [1989]. Evolution and the generative order. In: Goodwin, Brian; Saunders, Peter (eds.), *Theoretical Biology: Epigenetic and Evolutionary Order from Complex Systems*. Baltimore: Johns Hopkins University Press, 89–100.
Gould, Stephen Jay 2002. *The Structure of Evolutionary Theory*. Cambridge: Harvard University Press.
Gutmann, Mathias 2004. Uexküll and contemporary biology: Some methodological reconsiderations. *Sign Systems Studies* 32(1/2): 169–186.
Harnad, Stevan 1990. The symbol grounding problem. *Physica* D 42: 335–346.
Hawkes, Terence 2005 [1977]. *Structuralism and Semiotics*. 2nd ed. London: Routledge.
Hoffmeyer, Jesper 1996. *Signs of Meaning in the Universe*. Bloomington: Indiana University Press.
Hoffmeyer, Jesper 2008a. *Biosemiotics: An Examination into the Signs of Life and the Life of Signs*. Scranton: Scranton University Press.

Hoffmeyer, Jesper (ed.) 2008b. *A Legacy for Living Systems: Gregory Bateson as Precursor to Biosemiotics.* Berlin: Springer.
Hoffmeyer, Jesper 2009. Biology is immature biosemiotics. In: Deely, John; Sbrocchi, Leonard G. (eds.), *Semiotics 2008: Specialization, Semiosis, Semiotics.* (Proceedings of the 33rd Annual Meeting of the Semiotic Society of America, Houston, TX, 16–19 October 2008.) Ottawa: Legas, 927–942.
Jakobson, Roman 1971 [1929]. *Selected Writings: Vol. 2. Word and Language.* The Hague: Mouton.
Jiménez-Montano, Miguel A. 1992 [1989]. Formal languages and theoretical molecular biology. In: Goodwin, Brian; Saunders, Peter (eds.), *Theoretical Biology: Epigenetic and Evolutionary Order from Complex Systems.* Baltimore: Johns Hopkins University Press, 199–210.
Krampen, Martin 1997. Models of semiosis. In: Posner, Roland; Robering, Klaus; Sebeok, Thomas A. (eds.), *Semiotics: A Handbook on the Sign-Theoretic Foundations of Nature and Culture*, vol. 1. Berlin: Walter de Gruyter, 247–287.
Krois, John 2004. Ernst Cassirer's philosophy of biology. *Sign Systems Studies* 32(1/2): 277–295.
Kull, Kalevi 1992. Evolution and semiotics. In: Sebeok, Thomas A.; Umiker-Sebeok, Jean (eds.), *Biosemiotics: Semiotic Web 1991.* Berlin: Mouton de Gruyter, 221–233.
Kull, Kalevi 1999. Biosemiotics in the twentieth century: A view from biology. *Semiotica* 127(1/4): 385–414.
Kull, Kalevi 2001. A note on biorhetorics. *Sign Systems Studies* 29(2): 693–704.
Kull, Kalevi 2003. Ladder, tree, web: The ages of biological understanding. *Sign Systems Studies* 31(2): 589–603.
Kull, Kalevi 2009. Biosemiotics: To know, what life knows. *Cybernetics and Human Knowing* 16(3/4): 81–88.
Lévi-Strauss, Claude 1981. *The Naked Man.* (Weightman, John; Weightman, Doreen, trans.) New York: Harper & Row.
Maran, Timo 2008. *Mimikri semiootika.* Tartu: Tartu University Press.
Markoš, Anton 2000. *Tajemství hladiny: Hermeneutika živého.* Praha: Vesmír.
Markoš, Anton 2002. *Readers of the Book of Life: Contextualizing Developmental Evolutionary Biology.* Oxford: Oxford University Press.
Martinelli, Dario 2007. *Zoosemiotics: Proposals for a Handbook.* (Acta Semiotica Fennica 26). Imatra: International Semiotics Institute.
Meyer-Abich, Adolf 1963. *Geistesgeschichtliche Grundlagen der Biologie.* Stuttgart: G. Fischer Verlag.
Merrell, Floyd 1996. *Signs Grow: Semiosis and Life Processes.* Toronto: University of Toronto Press.
Millikan, Ruth G. 1989. Biosemantics. *Journal of Philosophy* 86: 281–297.
Mounin, Georges 1970. *Introduction à la sémiologie.* Paris: Éditions de Minuit.
Neuman, Yair 2006. Why do we need signs in biology. *Rivista di Biologia/Biologia Forum* 98: 497–513.

Neuman, Yair 2008. *Reviving the living: Meaning making in living systems.* New York: Elsevier.
Niklas, Ursula 1991. Semiotics and hermeneutics. In: Sebeok, Thomas A.; Umiker-Sebeok, Jean (eds.), *Recent Developments in Theory and History: The Semiotic Web 1990.* Berlin: Mouton de Gruyter, 267–283.
Nöth, Winfried 1990. *Handbook of Semiotics.* Bloomington: Indiana University Press.
Pain, Stephen 2002. Biorhetorics: An introduction to applied rhetoric. *Sign Systems Studies* 30(2): 755–772.
Pain, Stephen 2009. From biorhetorics to zoorhetorics. *Sign Systems Studies* 37(3/4): 498–508.
Posner, Roland; Robering, Klaus; Sebeok, Thomas A. (eds.) 1997. *Semiotik: ein Handbuch zu den zeichentheoretischen Grundlagen von Natur und Kultur. [Semiotics: a handbook on the sign-theoretic foundations of nature and culture]* vol. 1. Berlin: Walter de Gruyter.
Rosen, Robert 1991. *Life Itself: A Comprehensive Inquiry into the Nature, Origin, and Fabrication of Life.* New York: Columbia University Press.
Rothschild, Friedrich Salomon 1962. Laws of symbolic mediation in the dynamics of self and personality. *Annals of the New York Academy of Sciences* 96: 774–784.
Rothschild, Friedrich Salomon 1968. Concepts and methods of biosemiotic. *Scripta Hierosolymitana* 20: 163–194.
Schogt, Henry G. 1994. Structuralism. In: Sebeok, Thomas A. (ed.), *Encyclopedic Dictionary of Semiotics*, 2nd ed. Berlin: Mouton de Gruyter, 980–984.
Schult, Joachim (ed.) 2004. *Biosemiotik — praktische Anwendung und Konsequenzen für die Einzelwissenschaften. (Studien zur Theorie der Biologie 6).* Berlin: Verlag für Wissenschaft und Bildung.
Sebeok, Thomas A. 1988. Communication, language and speech: Evolutionary considerations. In: Herzfeld, Michael; Melazzo, Lucio (eds.), *Semiotic Theory and Practice: Proceedings of the Third International Congress of the IASS Palermo, 1984*, vol. II. Berlin: Mouton de Gruyter, 1083–1091.
Sebeok, Thomas A. 1990. *Essays in Zoosemiotics.* Toronto: Toronto Semiotic Circle.
Sebeok, Thomas A.; Umiker-Sebeok, Jean (eds.) 1992. *Biosemiotics. The Semiotic Web 1991.* Berlin: Mouton de Gruyter.
Sériot, Patrick 1995. Lingvistika i biologiya. U istokov strukturalizma: biologicheskaya diskussiya v Rossii. [Linguistics and biology. At the sources of structuralism: biologists' controversy in Russia.] In: Stepanov, Yuri S. (ed.), *Yazyk i nauka kontsa 20 veka.* Moskva: RGGU, 321–341.
Sériot, Patrick 2001. *Struktura i tselostnost': Ob intellektual'nyh istokah strukturalizma v tsentral'noj i vostochnoj Evrope 1920–30-e gg.* Moskva: Yazyki Slavyanskoj Kul'tury.
Sériot, Patrick 2003. La pensée nomogénétique en URSS dans l'entre-deux-guerres: l'histoire d'un contre-programme. *Cahiers de l'ILSL* 14: 183–191.

Sermonti, Giuseppe; Sibatani, Atuhiro 1998. The "Osaka Group", ten years later. *Rivista di Biologia* 91: 125–158.
Smith, Jonathan Z. 2004. *Relating Religion: Essays in the Study of Religion*. Chicago: The University of Chicago Press.
Solé, Ricard; Goodwin, Brian 2000. *Signs of Life: How Complexity Pervades Biology*. New York: Basic Books.
Stepanov, Juri S. 1971. *Semiotika*. Moskva: Nauka.
Stjernfelt, Frederik 2007. *Diagrammatology: An Investigation on the Borderlines of Phenomenology, Ontology, and Semiotics*. Dordrecht: Springer.
Thom, René 1972. Structuralism and biology. In: Waddington, Conrad Hal (ed.), *Towards a Theoretical Biology*, vol. 4, *Essays*. Edinburgh: Edinburgh University Press, 68–82.
Thompson, D'Arcy Wentworth 1942 [1917]. *On Growth and Form*. 2nd ed. Cambridge: Cambridge University Press.
Uexküll, Jakob von 1930. *Die Lebenslehre*. Potsdam: Müller und Kiepenheuer Verlag.
Vetrov, Anatoli A. 1968. *Semiotika i ee osnovnye problemy*. Moskva: Izdatelstvo politicheskoj literatury.
Weber, Andreas 2003. *Natur als Bedeutung: Versuch einer semiotischen Theorie des Lebendigen*. Würzburg: Königshausen & Neumann.
Webster, Gerry 1989. Structuralism and Darwinism: Concepts for the study of form. In: Goodwin, Brian; Sibatani, Atuhiro; Webster, Gerry (eds.), *Dynamic Structures in Biology*. Edinburgh: Edinburgh University Press, 1–15.
Webster, Gerry; Goodwin, Brian 1996. *Form and Transformation: Generative and Relational Principles in Biology*. Cambridge: Cambridge University Press.
Wheeler, Wendy 2006. *The Whole Creature: Complexity, Biosemiotics and the Evolution of Culture*. London: Lawrence & Wishart.
Witzany, Günther 1993. *Natur der Sprache — Sprache der Natur: Sprachpragmatische Philosophie der Biologie*. Würzburg: Königshausen & Neumann.
Witzany, Günther (ed.) 2007. *Biosemiotics in Transdisciplinary Contexts: Proceedings of the Gathering in Biosemiotics 6, Salzburg 2006*. Salzburg: Umweb.
Zarenkov, Nikolai A. 2007. *Semioticheskaya teoriya biologicheskoj zhizni*. [Semiotic theory of biological life.] Moskva: KomKniga.

Part I
Biosemiotic Approach: General Principles

Chapter 2
Theses on Biosemiotics: Prolegomena to a Theoretical Biology

Kalevi Kull, Terrence Deacon, Claus Emmeche, Jesper Hoffmeyer and Frederik Stjernfelt

Summary. Theses on the semiotic study of life as presented here provide a collectively formulated set of statements on what biology needs to be focused on in order to describe life as a process based on semiosis, or sign action. An aim of the biosemiotic approach is to explain how life evolves through all varieties of forms of communication and signification (including the cellular adaptive behavior, animal communication, and human intellect), and to provide tools for grounding the sign theories. We introduce the concept of a semiotic threshold zone, and analyse the concepts of semiosis, function, umwelt, etc. as the basic concepts for theoretical biology.

The variety of scientific disciplines that constitute modern biology, and the science of sign systems, semiotics, commonly known from the study of human language and social signs systems, have recently demonstrated trends towards a recognition that sign processes *per se* and the processes of life may be intimately and inseparably interconnected. This view has developed into a general approach called biosemiotics. The name "biosemiotics" appears to have been coined by Friedrich S. Rothschild in 1962, but Thomas Sebeok played a major role in defining the field in the 1980s and 1990s (e.g., Anderson *et al.* 1984; Sebeok, Umiker-Sebeok 1992; Sebeok 1996; 2001).[1]

---

[1] On the history of biosemiotics, see Favareau (2007); Kull (1999).

There have been several attempts to formulate the basic principles of biosemiotics, both in extensive (Hoffmeyer 1996; 2008) and more compact versions (Sebeok 1996; Hoffmeyer 1997; Emmeche et al. 2002; Stjernfelt 2002), together with detailed analyses of some central biosemiotic problems (Deacon 1997; Kull et al. 2008). Still, the establishment of biosemiotics requires, on the one hand, a deepening and grounding of the theory of semiotics, and on the other hand, a development of a richer theoretical biology. But as those of us who identify as biosemioticians have begun to organize international associations, annual meetings, edited volumes, and journals devoted to this new field, it has become apparent that a single well-defined paradigm is still in the process of coalescing from a diverse collection of theoretical positions. Although such diversity is a healthy starting point from which to develop an intellectually productive field of research, it is also important to develop a clear sense of the scope of these various visions of the field. To accomplish this it is necessary to first identify points of common terminology and shared theoretical assumptions, and then to identify incompatible frameworks and conceptual issues that still need to be resolved. This document represents an effort to articulate a common set of assumptions that are shared among a group of researchers in the field who ground their work on a Peircean framework. It is hoped that by carefully outlining and analyzing our shared theoretical assumptions, this will help clarify and contrast this approach with respect to others.

Because a significant number of those of us who identify as Peircean biosemioticians are located in northern Europe, we agreed to meet and attempt to formulate such a document during August 1–6, 2008 at a meeting in Estonia (at Saka cliff on the coast of the Baltic Sea) by invitation of Tartu semioticians. Our aim was to formulate a joint understanding of the conceptual basis and basic principles of a semiotic study of life, i.e., biosemiotics. The results are arranged in the form of the eight theses that follow.

The eight theses

I. *The semiosic/non-semiosic distinction is co-extensive with life/non-life distinction, i.e., with the domain of general biology*

The concepts of *function* and *semiosis* (sign processes) are intertwined. Both are teleological concepts in the sense of being determined with respect to an end (or *other* than itself) — a specifically correlated *absent* content. Although it is unclear whether these two properties of living processes (function and semiosis) are exactly co-extensive, it is clear that although time-asymmetrical, irreversible physical processes are found in the prebiotic physical–chemical world, teleological processes that are specially organized with respect to specific ends or referents are unique to living processes.

If we conceive of a function as a process organized around an implicitly represented end, then these two classes of phenomena must be considered entirely co-extensive. Alternatively, semiosis, the activity of sign processes, may be considered only in conditions where there is explicit or implicit representation of an end state or where a functional satisfaction condition can be identified as holding or not holding, in which case semiosis can be defined with respect to prior function.

With the demonstration of the plausibility of proto-life processes, such as autocells (Deacon 2006a), which are recursively self-maintenant, and so can self-repair, reproduce, and evolve, it becomes difficult to identify the exact threshold of the onset of semiosis, and yet function can be unambiguously demonstrated by the presence of the component processes. This, then, allows us to identify a lower limit to the identification of functional organization — thus *telos* — even though explicit semiosis is still ambiguous. This identifies a *threshold zone* below which semiosis is not defined but above which there can be stages of semiotic differentiation. It is thus an open and crucial issue of research to determine, empirically and conceptually, the different thresholds in this zone between such simple reproducing and evolving systems and contemporary terrestrial organisms that appear to depend unambiguously on semiotic processes.

Often the emergence of life is seen as a sudden transition where the many properties defining life arise together or are tightly interconnected (like self-replication, autocatalysis, function, cellularity, etc.). However, this appears to be both too simple and too implausible. There is not simply one dividing line where all the interconnected properties of living systems, as we know them, emerge. Instead we observe what we call a *threshold zone*, likely involving incremental stages in which different component processes emerge. This is an open issue for further investigation, and will likely develop into a fertile area for both molecular biology and biosemiotic research to contribute.

II. *Biology is incomplete as a science in the absence of explicit semiotic grounding*

The neodarwinian biology as practiced all over the world has prescinded (i.e., abstracted from necessary contextual support) an asemiotic conception of life as mere molecular chemistry, and yet at the same time is dependent on unanalyzed semiotic assumptions. The reason why this is not felt as a problem is that biology compensates for the excluded semiosis by introducing a plethora of implicitly semiotic terms like "information", "adaptation", "signal", "cue", "code", "messenger", "fidelity", "cross talk", etc. These uses are seldom well defined, and are often applied in an allegedly metaphoric way, with the implicit assumption that they can be reduced to mere chemical accounts if necessary.

It is not clear, however, that a complete and unproblematic reduction of this sort is possible. If biologists were asked to avoid these implicitly semiotic terms they would have a hard — and probably impossible — job of explaining the nature of organic function. For example, if hemoglobin was known only by its 3D molecular structure it would not be possible to guess that it functioned as a transporter for oxygen. But knowing that hemoglobin is a reflection of the need of multicellular organisms to provide energy for the metabolism of somatic tissues, it immediately becomes clear (1) that it must have some structural features conducive to binding and transporting oxygen in blood; (2) that the oxygen-binding region of the hemoglobin molecule is expected to be conserved throughout evolution; (3) that different forms of hemoglobin differ in specific ways that correspond to different oxygen transport requirements (e.g., in different species or in mammalian gestation).

The theoretical issue at stake here is that in biology empirical facts are always contextually constrained. Contextuality should not be conceived as a free ticket to determinations from outside domains — rather, contextuality is constrained by function (and vice versa). If life exists on distant planets in other regions of the universe, with large heterotrophic forms it is likely that they will also require a corresponding transporter molecule, and if their metabolism is mediated by oxygen then the transporter molecules might not resemble our Earth-hemoglobin in narrow molecular detail, but would none the less retain the capacity to carry oxygen in a similarly protected way so that it can be released again in tissues in need of it. And if their metabolism is mediated by another substance than oxygen, they will still require some substance undertaking the analogous function of facilitating energy transport. This already puts constraints upon the space of possible realizations of the "hemoglobin molecule" (or any such nonterrestrial functional analog), which severely restricts its form and its correspondence to contextual factors.

Another way to put this is to say that hemoglobin function is not intrinsic to its molecular structure. Rather it is relational — hemoglobin may be seen as a carrier of *constitutive absence* (Deacon 2006b), in the sense that the molecule's properties are constituted not only by intrinsic features, but by extrinsic features of its historical and physical functional contexts. In effect, the missing oxygen with respect to which hemoglobin structure has evolved has become its defining characteristic. In this respect, one can understand the structure of hemoglobin as a "representation" of both oxygen and its role in the cellular molecular processes of metabolism. The function of hemoglobin is in this way also what affords the possibility of it having representational character. This function relates to the "needs" or self-maintenance conditions of some agent. Needing something implies both its transient absence and some structure or processual state representing that absence and its possible ending or completion.

This constitution with respect to something extrinsic and/or absent shows that function and representation are two aspects of the same mode of relational existence. This implies that the primary unit of biosemiotic research is a sign, not merely a molecule or cell.

III. *The predictive power of biology is embedded in the functional aspect and cannot be based on chemistry alone*

It is an accepted truth in biology that structure and function are interdependent, e.g., a biological explanation is incomplete even if you have exhaustively described the production and the structure of a macromolecule in a cell. There is still a missing feature of the explanation — we still need an answer to the question: "what is it for?" Answering this question is part of the functional contextualization that all biological facts require. In many cases, these functions are characterized as being of a regulatory, information-carrying or signaling kind, and thus describing the function of these structures is really embedding them in a wider system that has a sign-processing (e.g., signal-transducing) character. Here, in this wider system, the functions that a macromolecule take part in (or contribute to) can be fulfilled by other slightly or completely different structures (say, being only similar regarding an "active site") and thus, the structure is seen as a vehicle for fulfilling that function. Locating and clarifying functional–processual parts in a wider network within metabolism enables partial predictions of some constraints needed to be met if the vehicle of a function should be able to work. Most of the predictive power of biology is lost if semio-functional analysis is excluded.

The difficulty of making predictions about biological phenomena is that the functions are plurally realizable and thus subject to considerable variation. As a result, the physical–chemical details necessarily provide an incomplete account. Functional requirements do, however, constrain the physical–chemical substrates that can be recruited.

IV. *Differences in methodology distinguish a semiotic biology from the non-semiotic one*

It is the aim of biosemiotics to make explicit those assumptions that are imported into biology by such unanalyzed teleological concepts as "function", "information", "code", "signal", "cue", etc., and to provide a theoretical grounding for these concepts. The widespread use of such terms in existing biology points to the fact that such notions cannot be avoided or fully substituted with merely chemical accounts. Biosemiotics has the scientific tasks of (1) grounding such terms in a physico-biological context; (2) defining and interrelating such terms with the constant aim of avoiding the anthropomorphisms which threaten when they are left with implicit definitions only; and (3) to make biology theoretically complete.

One of the early masterpieces of theoretical biology, Kant's "Kritik der Urteilskraft", provides some interesting definitions of teleology in biology. He describes organisms as possessing a "formative power" to construct themselves as an "organized and self-organized being" in which "every part is reciprocally both ends and means" (Kant 1790: Part II, paragraph 66; see also Stjernfelt 2007: 199ff). It is important to underline that in making these definitions of teleology, Kant is, as so often, performing a criticism of naive metaphysics — where the idea of a *telos* is treated as something external to nature which by means of strange, unknown powers affects natural processes. By contrast, this definition of *telos* is functional and thus *internal* to nature and characterizing a specific class of natural processes. It is a concept of *telos* that does not refer to unknown vitalist forces, but rather defines *telos* by a specific class of causal processes.

Modern biology has been working on the assumption that there is an incompatibility between these teleological and physical–chemical characterizations of life. Biosemiotic approaches assume that there is no deep incompatibility and that a principled theory unifying these domains is possible.

V. *Function is intrinsically related to organization, signification, and the concept of an autonomous agent or self*

Functions are not only the output of evolutionary history, rather functionality is the prerequisite for organic evolution. For instance, autocells do not in all cases have an evolutionary history but they do have functions. Evolution presupposes function, rather than vice versa. Natural selection cannot be defined except with respect to a bounded self-maintaining and self-reproducing dynamical unit system. A discrete system with these properties must therefore be constituted by component materials and dynamical processes that reciprocally generate each other and their collective organization. The critical features and dynamical actions of these components exemplify Kant's criteria for possessing intrinsic *telos* and are thus functional. The possibility of evolution derives from the fact that functions, because they can be multiply realized, can coopt any incidental physical–chemical properties of the substrates they utilize. Semiosis, likewise, can coopt any incidental features exhibited by functional processes or their properties.

An unresolved question arises with respect to the relationship between semiosis and function, and which terminological paradigms can best characterize this fundamental relationship. Thus, the coupling of perception signs and action signs in Uexküll's functional cycle (Uexküll 1921; Kull 2001) has an *if–then* inference structure.[2] Perception signs form the premise of the conclusion inherent in the ensuing action signs and grant (with some probability) that something is the case, and consequently, the organism "makes a decision" to act on the basis of this information. This functional cycle thus has an *if–then* structure, like a fallible inference of either an inductive, deductive or abductive form. However, defining functional and semiotic processes by reference to the concept of inference risks the charge of circularity. To avoid confusion, other terms might be preferred: e.g., conditional causality, proto-argument, or the like.

Peirce identified semiotics with logic in the broadest sense. As we have implied above, adaptation also involves the selective semiotic recruitment of those physical–chemical aspects of the organism-environment that are relevant to the persistence of that process. As a consequence, conditional relationships of logic become re-presented in the forms and habits of organisms and their components embodying this bio-logic. In this way, semiosis facilitates the development of an organism's capacity to behave in a way that is both consistent with its environment and implicitly inferential. "Logic" as we are using it here is not something to be considered as a product of abstract cognition in humans, but rather we simply intend to highlight the inference-like architecture of biological function, which we take to also be the basis of semiosis in general.

VI. *The grounding of general semiotics has to use biosemiotic tools*

Biosemiotics does not take for granted the wide variety of concepts of the sign, sign action, etc., in the different semiotic traditions, but

---

[2] This might align it with Peirce's notion of an argument; "argument" here is of course not taken in the sense of symbolic logic where all internal structure of it needs to be explicit, but in a broader Peircean sense as signs whose interpretants — here the resultative action signs — are made explicit.

undogmatically sees these as a resource for the construction of an up-to-date, refined, and better-grounded (as concerns contemporary biology) version of a general semiotics. The aim is to understand the dynamics of organic mechanisms for the emergence of semiotic functions, in a way that is compatible with the findings of contemporary biology and yet also reflects the developmental and evolutionary history of sign functions.

What do signs do? They stabilize or secure reliable modes of self-maintenance in such a way that they are able to expand the realm of processes that have already proved functional in the past.[3] They do this in an economical way, allowing the recognition of no more than an aspect of an object to suffice for the organism to act upon that object. Of course, the flip side of this economy is the possibility of fallibility, but fallibility is also what provides the space of alternatives that makes evolution possible. This securing of prior forms and dynamical relationships implies "remembering" what has already proved functional for self-maintenance. Remembering a bio-form is like remembering its recipe for its way of production, or regeneration.[4] To remember in this biological sense is to be able (for some sign system) to put to use that set of constraints or imposition of boundary conditions that confine physical and chemical processes to actualize the means of production of these same forms.[5]

---

[3]Hoffmeyer introduced the term *semiotic scaffolding* in order to characterize the role of sign processes: each step in ontogeny is temporarily supported by a web of internal sign processes assuring the correct direction of the process. A similar supportive semiotic interaction structure (at the ecological level) may play a role in phylogeny.

[4]The form of a process is not its actual existence but its participation in a general class of processes, and this class can reciprocally be defined with respect to a common functional consequence achieved.

[5]Confer Peirce: "That which is communicated from the Object through the Sign to the Interpretant is a Form; that is to say, it is nothing like an existent, but is a power, is the fact that something would happen under certain conditions" (Peirce, MS 793: 1–3; cf. EP2, 544, n. 22).

VII. *Semiosis is a central concept for biology — however, it requires a more exact definition*

Although there are many descriptions of semiotic processes, it is still an unresolved challenge to provide an account that explains what exactly constitutes semiosis without either assuming a homuncular interpreter or else leaving critical relationships undefined. While this is not so problematic for human or complex animal communication, where an interpreter can be provisionally assumed without further explanation, it becomes a serious challenge for fundamental issues in biosemiotics, since we cannot in these cases appeal to an extrinsic interpreter. The organism (or the organism plus its environment) must, in itself, constitute an interpreter, but in biosemiotic analysis we must attempt to be explicit in explaining specifically which processes provide the necessary and sufficient conditions to consider that process semiosis.

The interpretive capacity is an emergent property of a reciprocal end-means relationship of a self-propagating dynamical system. The constitutive absence (Deacon 2006b, discussed above) is the basis of both biological function to and dependence on an environment. Because an organism must incessantly remake itself utilizing resources afforded by its environment, it must be in dynamical correspondence with these crucial intrinsically absent features, and at the same time its constituent parts and dynamics must be reciprocally generating one another with respect to this absence. In this respect, an organism is a sign-interpreting process that can be described as a recursive self-referential sign production process, dependent on or influenced by some external factors likely to be present in its environment.

We are not currently in a position to provide a more precise and unambiguous description of the interpretive architecture that is implicit in an organism. However, we can identify many critical component processes and relationships that must be involved, and can provide a rough sketch of what a most simple model of the creation of a semiotic relationship should involve.

We can identify seven properties or conditions that must be met. The following is a rough sketch of these critical conditions.

1. *Agency*: A unit system with the capacity to generate end-directed behaviors.

2. *Normativity*: A semiotic process builds up normative properties in a broad sense, thus being itself embedded in a process that contributes normativity. This includes the possibility that the representation is in error or that its consequence (in Peircean terms: its dynamical interpretant) can be either compatible or incompatible with preserving the integrity of the living system in which it occurs.

To grasp this minimal notion of normativity, think of the difference between a physical pattern as such and that pattern serving a function. Any specific physical pattern may be characterized by algorithmic information theory as either highly random, or highly regular, or something complex in between, be it either descriptively compressible or truly complex and incompressible. But a pattern serving a function has, in addition to its own high or low algorithmic information content, a degree to which it serves, or fails to serve, its goal. For semiotic processes (having such functions as representation, information storing, and interpretation) the degree to which a pattern serves or fails to serve such functions constitutes a norm.

3. *Teleo-functionality*: Semiosis is always embedded in a process that is end-directed in which the semiosis can be assessed with respect to whether its interpretation is concordant or discordant with the dynamics of achieving that end. This is what determines the normative properties of a sign-interpreting process.

4. *Form generation*: The systemic organization that is responsible for interpreting the semiotic function of a sign vehicle must include a form-generating process that in an either direct or indirect way contributes to the persistence (re-presentation) of that function. The

interpretation process is constituted by generating a structure (physical form) that serves as a sign of the prior sign and also can produce further structural consequences.

5. *Differentiation of a sign vehicle* from the dynamics of the reciprocal form-generating process: A sign vehicle must be insulated from the dynamics that it constrains and that is responsible for generating a repetition of this process.[6]

6. *Categorization*: Sign repetition can never be 100 per cent physically identical. This, together with normativity, is why signs form types. Categorization appears in all communication processes in case of adaptive-enough systems. Functionally similar instances of signs (tokens) are subsumed under a general type. At the same time, what signs refer to is also categorized. This aspect of sign use is highly economical because it enables the organism to get by with the generation of only a finite number of simple, typical modes of interpretive action to achieve similar ends. The flip side of this is, of course, the possibility of fallacy, cf. above.

7. *Inheritance of relations*: Various developmental processes are those that create novel fitted correspondence relationships among the parts of the organism and among the organism and its environment, and are presumed to be the principal means by which semiotic relationships are generated. In this respect, genetic inheritance represents one of the most basic forms of semiosis, and so studying the conditions of its generation should provide insight into the way

---

[6]This is what H. H. Pattee has pointed out many times (e.g., Pattee 2007). We think that Pattee shares with us the aim of seeing the distinction between a sign vehicle and its dynamics as a product of evolution, and not simply as a taken-for-granted primitive irreducible distinction between matter and symbol (*pace* his own terminology). The mature Peircean notion of a sign (and Peirce's developmental taxonomies of inclusive, specialized and degenerative types of sign aspects) is probably a more fruitful point of departure than common sense, linguist or computer-science notions of symbols.

semiotic process become grounded in physical processes. Besides genetic inheritance, there are several other forms of inheritance (e.g., epigenetic, neural, social, etc.) that are in use by various communication processes. In this sense, semiotic processes include memory processes in general that maintain continuity of information and stability of dynamical options.

Thus, *aboutness* can exist without invoking mental (*sensu stricto*) operations (processes taking place in brains, possibly involving, e.g., awareness, consciousness). Conversely, mental operations in this sense may evolve as a higher-order augmentation of the capacities to generate and process aboutness.

VIII. *Organisms create their umwelten*

Organisms, as embodiments of semioses, are not separable from the environment without losing their essential nature. Therefore, a number of specific concepts that describe these relationships are essential.

The *umwelt* is the set of features of the environment as distinguished by the organism, or the self-centered world that relates an organism with everything else. This concept was introduced into biology by Jakob von Uexküll and became widely used and further developed in semiotics, anthropology, philosophy and elsewhere, especially since the late 1970s (Uexküll 1921; 1982 [1940]).

A *semiotic niche* is defined as the totality of signs or cues in the surroundings of an organism — signs that it must be able to meaningfully interpret to ensure its balance and welfare. The semiotic niche includes the traditional ecological niche factors, but now the semiotic dimension of these factors is also emphasized. The organism must *distinguish* relevant from irrelevant food items and threats, for example, and it must *identify* the necessary markers of the biotic and abiotic resources it needs: water, shelter, nest-building materials, mating partners, etc. The semiotic niche thus comprises all the *interpretive challenges* that the ecological niche forces upon a species.

The semiotic niche in this way may be seen as an *externalistic counterpart* to the *umwelt* concept: if the umwelt denotes an internal model in the organism, then the semiotic niche refers to a segment of the external environment. It makes the umwelt concept compatible with an evolutionary approach, since now one may pose the question of whether the umwelt of a species is sufficiently differentiated to meet the challenges posed by the available semiotic-niche conditions.

Conclusions

Biosemiotics sees itself as an extended and more general approach to biological explanation that complements and augments the concept of biological function. Thus the physical–chemical account of biological phenomena can be seen as a special case within biosemiotics. Likewise, this concerns the different branches and theories in biology, for instance the neodarwinian theory of evolution occurs as a special case of a biosemiotic theory of evolution.

In relation to semiotics, biosemiotics provides a way for grounding its theory. To the extent that semiotic theories of social and communicative processes at the human scale are defined with respect to human or animal interpreters, and thus defined indirectly with respect to the biological processes constituting minds, most semiotic theories must implicitly appeal to a biosemiotic interpretation process.

Recognizing this necessary dependence is the first step toward understanding how the humanities and sciences might be integrated into a new grand synthetic theory without having to reduce one to the other.

References

Anderson, Myrdene; Deely, John; Krampen, Martin; Ransdell, Joseph; Sebeok, Thomas A.; Uexküll, Thure von 1984. A semiotic perspective on the sciences: Steps toward a new paradigm. *Semiotica* 52(1/2): 7–47.

Deacon, Terrence 1997. *The Symbolic Species*. London: Penguin.
Deacon, Terrence 2006a. Reciprocal linkage between self-organizing processes is sufficient for self-reproduction and evolvability. *Biological Theory* 1(2): 136–149.
Deacon, Terrence 2006b. Emergence: The hole at the wheel's hub. In: Clayton, Philip; Davies, Paul (eds.), *The Re-Emergence of Emergence*. Oxford: Oxford University Press, 111–150.
Emmeche, Claus 2002. The chicken and the Orphean egg: On the function of meaning and the meaning of function. *Sign Systems Studies* 30(1): 15–32.
Emmeche, Claus; Kull, Kalevi; Stjernfelt, Frederik 2002. A biosemiotic building: 13 theses. In: Emmeche, Claus; Kull, Kalevi; Stjernfelt, Frederik, *Reading Hoffmeyer, Rethinking Biology*. (Tartu Semiotics Library 3.) Tartu: Tartu University Press, 13–24.
Favareau, Donald 2007. The evolutionary history of biosemiotics. In: Barbieri, Marcello (ed.), *Introduction to Biosemiotics: The New Biological Synthesis*. Dordrecht: Springer, 1–67.
Hoffmeyer, Jesper 1996. *Signs of Meaning in the Universe*. Bloomington: Indiana University Press.
Hoffmeyer, Jesper 1997. Biosemiotics: Towards a new synthesis in biology. *European Journal for Semiotic Studies* 9(2): 355–376.
Hoffmeyer, Jesper 2008. *Biosemiotics: An Examination into the Signs of Life and the Life of Signs*. Scranton: Scranton University Press.
Kant, Immanuel 1952 [1790]. *The Critique of Judgement*. [*Kritik der Urteilskraft*.] (Meredith, James Creed, trans.) Oxford: Oxford University Press.
Kull, Kalevi 1999. Biosemiotics in the twentieth century: A view from biology. *Semiotica* 127(1–4): 385–414.
Kull, Kalevi (ed.) 2001. Jakob von Uexküll: A paradigm for biology and semiotics. *Semiotica* 134(1–4): 1–828.
Kull, Kalevi; Emmeche, Claus; Favareau, Donald 2008. Biosemiotic questions. *Biosemiotics* 1(1): 41–55.
Pattee, Howard H. 2007. The necessity of biosemiotics: Matter-symbol complementarity. In: Barbieri, Marcello (ed.), *Introduction to Biosemiotics: The New Biological Synthesis*. Dordrecht: Springer, 115–132.
Peirce, Charles Sanders 1931–1935, 1958. *The Collected Papers of Charles Sanders Peirce*. Vols. I–VI (Hartshorne, Charles; Weiss, Paul, eds.); vols. VII–VIII (Burks, Arthur W., ed.). Cambridge: Harvard University Press. [Here referred to as CP, followed by volume and paragraph number.]
Peirce, Charles Sanders 1998 [1893–1913]. *The Essential Peirce: Selected Philosophical Writings*. Vol. II. (Peirce Edition Project, ed.) Bloomington: Indiana University Press. [Here referred to as EP2, followed by the number of the page.]
Rothschild, Friedrich Salomon 1962. Laws of symbolic mediation in the dynamics of self and personality. *Annals of New York Academy of Sciences* 96: 774–784.

Sebeok, Thomas A. 1996. Signs, bridges, origins. In: Trabant, Jürgen (ed.), *Origins of Language*. Budapest: Collegium Budapest, 89–115.

Sebeok, Thomas A. 2001. Biosemiotics: Its roots, proliferation, and prospects. *Semiotica* 134(1–4): 61–78.

Sebeok, Thomas A.; Umiker-Sebeok, Jean (eds.) 1992. *Biosemiotics: The Semiotic Web 1991*. Berlin: Mouton de Gruyter.

Stjernfelt, Frederik 2002. Tractatus Hoffmeyerensis: Biosemiotics as expressed in 22 basic hypotheses. *Sign Systems Studies* 30(1): 337–345.

Stjernfelt, Frederik 2007. *Diagrammatology: An Investigation on the Borderlines of Phenomenology, Ontology, and Semiotics*. Dordrecht: Springer Verlag.

Uexküll, Jakob von 1921. *Umwelt und Innenwelt der Tiere*. 2te Auflage. Berlin: Julius Springer.

Uexküll, Jakob von 1982 [1940]. The theory of meaning. *Semiotica* 42(1): 25–82.

Chapter 3
Biology Is Immature Biosemiotics

Jesper Hoffmeyer

Summary. Biology as practised in laboratory or field studies all over the world has prescinded the asemiotic basics of life from its full expression. The reason why this is not felt as a problem is that biology compensates for the excluded semiosis by introducing a whole group of implicitly semiotic terms like information, adaptation, signal, messenger, fidelity, cross-talk etc. If biology was asked to avoid these implicitly semiotic terms it would have a hard — and probably impossible — job of explaining function. The theoretical issue at stake here is that in biology empirical facts are always contextually constrained. A full understanding of structural or behavioral elements implies that the "what for" question is also posed, but since function is an end-directed activity an asemiotic explanation for its occurrence leads us into mystery (or hidden homunculi). The relation between semiotic causation and efficient causation will be discussed.

Function in biology

One reason why biologists, and scientists in general, so vehemently oppose any claim to the effect that natural selection might not be a sufficient explanation for evolution on Earth, is the belief that the theory of natural selection is the very element that glues biology to materialistic science. The disquieting fact that the creatures of this world so clearly exhibit purposeful behavior has always tempted philosophically minded biologists to claim the existence of peculiar vital forces or principles pertaining to life. It is probably no longer possible to make a career in biology at the university level if you adhere to such vitalist ideas, but biologists still have to somehow cope with the obvious, though tabooed, teleological aspect of life. The prime mode in which this is done in modern biology is by the application of the concept of *function*. Volumes have been produced to solve the problem of how to justify the concept of function inside a non-teleological frame of understanding. And this is where natural selection comes in, for natural selection will tend to optimize the capacity of species to meet the functional challenges of their ecological niche conditions. Functionality is exactly what natural selection is supposed to produce.

The term "function" in biology is understood as the answer to a question about *why* some object or process has evolved in a system. In other words, what is it good for? A function thus refers forward in time from the object or process, along some chain of causation to the goal or success. This inversed arrow of time (future directedness) immediately sets functions apart from other kinds of mechanisms that always refer backward along some chain of causation explaining how the feature occurred. Darwinists, however, are not worried about the teleological character of functions because they believe that natural selection will ultimately account for them through ordinary mechanistic causation. Thus, as often noted by Darwinists, adaptive traits are *not* explained by the consequences they *will* or *can* have but by the consequences they already *have had* in ancestor populations. The consequences in other words precede the effect they explain, and selection does not therefore challenge the mechanistic paradigm of traditional biology. So, as the explanation goes, the teleology implied by the concept of function is only an "as if" teleology, i.e., a *teleonomy*.

But something feels wrong with this argument.[1] There is of course no reason to believe that the flowering apple tree outside my window right now has any purpose in producing flowers. Trees do

---

[1] Cf. a discussion of teleonomy in Deely (2001): 65–66.

not have purposes in the human sense of the term. But clearly even apple trees exhibit some kind of *agency* in the sense that they have a capacity to generate end-directed behaviors. The apple flowers serve to attract bees that can pollinate them and thus assure the occurrence of sexual reproduction. But how can a purely mechanical principle (natural selection) produce agency in a world that is not supposed to carry any trace of end-directedness? Our intuitive sense of logic seems to be strained quite a bit here.

We shall return to this problem, but let us first in some more detail consider a clear-cut case of a biological function, the heart: what is the function of the heart? Several answers might be given depending on the level at which the problem is answered. Thus, considering that the heart beats faster in states of emotional agitation one might claim that the function of the heart is to make us aware of states of emotional agitation. A more imaginative answer was put forward by complexity researcher Stuart Kauffman, who once observed that our heartbeat may reflect the outburst of an upcoming earthquake before it actually hits us. Although this early warning signal is not at present the function of the heart, one might speculate that if earthquakes were to become more frequent in the future survivors would tend to possess hearts that have optimized this function of being an early warning system. The beating heart would then have become an example of so-called Darwinian preadaptation (or exaptation [Gould, Vrba 1982]), a non-functional trait which under changed circumstances is recruited by natural selection for a whole new function, in this case a warning signal against imminent earthquakes.

But the main function of the heart is of course to pump blood. The heart is effectively nothing but a muscle, and in spite of much prize for its wisdom it doesn't even know we exist. It nevertheless serves us well from the moment it starts beating early in embryonic development, and continues to serve us by beating rhythmically 60–80 times per minute without any serious interruptions until the moment we die some 75 years later (on average). In a normal lifetime the heart will have beaten approximately three billion times; the heart therefore may well be dumb but it is certainly not lazy.

Pumping blood, however, is only the most proximal function of the heart; for in very small animals, where oxygen can penetrate into the cells from the surface of the animal, there is no need for a heart to pump oxygen carrying liquid around. Pumping blood therefore is only a vital function in animal life because it is a tool for the more central function of providing oxygen to the tissues. So, the real function of the heart, it might be said, is to provide oxygen. But even this function does not hit the "heart" of the matter, for there are plenty of organisms on Earth that do not need oxygen to drive their metabolism, and oxygen may even be toxic to such organisms. The reason oxygen is needed in animals is that the metabolism of animals happens to be based on the burning of food through a series of oxidative processes that use oxygen as the ultimate oxidative source. Therefore oxygen delivery as such is not the core of the matter, rather the core is that organisms need to assure the metabolic formation of energy-carrying molecules (ATP). So the primary function of the heart might be seen as that of assuring ATP production in animal tissues (Fig. 1).

Strictly speaking, however, even the function of providing a well-regulated source for the production of energy carrying molecules (ATP) cannot be said to be the ultimate function of the heart. For if an organism did not engage in reproduction, evolution would not care whether it carried a heart or not. Only because organisms need to reproduce is the heart of importance to evolution. So, seen in a wider perspective the real function of the heart is to make reproduction possible, and so on.

What this mapping of heart functions shows us is that biological functions are contextual and ultimately they depend on the simple fact that life is historical. Life is historical in the sense that its continuation depends on an ability to learn: strategies that have proved effective in overcoming past challenges must be "remembered" so that descendent organisms will be able to cope with those same

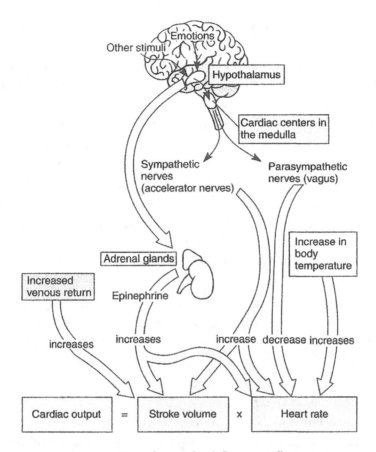

Some factors that influence cardiac output.

Fig. 1. The semiotics of heart function. Heart function is extremely well controlled by the animal body. A subtle semiotic web is in charge of this control. Sensory receptors in the walls of certain blood vessels and heart chambers are sensitive to changes in blood pressure. When stimulated they send messages to the cardiac centers in the medulla of the brain. The cardiac centers in the brain maintain control over two sets of autonomic nerves that pass to the SA node (or pacemaker): Parasympathetic nerves works through a single nerve — the vagus nerve — which release acetylcholine, which slows the heart; sympathetic nerves release norepinephrine which speeds heart rate and increases the strength of contraction. In response to physical or emotional stress the adrenal glands release epinephrine and norepinephrine, which speed up the heart rate. Psychological states interact through the hypothalamus with sympathetic nerves and — more indirectly — with parasympathetic nerves.

challenges; and — just as importantly — those strategies that did not work well must be "forgotten". The effect of these two processes is *learning*, and since natural selection is precisely a mechanism for remembering the fit and forgetting the unfit, natural selection is a kind of learning process. Now, learning processes are different from normal mechanical processes in that they depend on the formation of some form of a coded *representation* (the functional response must be incorporated in the system in some way or other, for instance as here in the form of the genetic setup). But the moment a representation becomes a constituent of a mechanism "misunderstandings" will necessarily lurk in the background.[2] An unpredictable source of change is thereby introduced into the system and change in systems with a capacity for learning is therefore historical in nature, not mechanical.

Learning is of course a semiotic process, an interpretative agency must be a part of the process, and in natural selection this interpretative agency resides in the lineage, i.e., the population as an evolving unit.[3] *Populations* are coterminous with the concrete individuals living here and now, while the *lineage* is a *historical* and *transgenerational subject* that possesses a *collective agency* as such, since its destiny as a temporal integrative structure is formed through the accumulated interactions of its single units (individuals) with their environments — much in the same way that multicellular organisms are integrative structures interacting with their surroundings via the activity of individual cells (Hoffmeyer 2008: 116). The lineage is the subject that may eventually learn to cope with changes in the ecological niche conditions or, otherwise, go extinct. According to present fashion in biology this learning process can be measured simply as

---

[2]Numerous diseases are known to interfere with normal heart function and some of them are caused by a "misreading" of important genes due to genetic deficiencies or to disturbances in normal embryonic development.

[3]The word *lineage* here is used in the conventional sense of a succession of ancestors and descendents, parents and offspring.

changes in gene frequencies down through the generations. Each generation in this simplified optic constitutes a cohort of genotypes expressing themselves as phenotypes that enter into the competitive acts of survival and reproduction. The resulting cohort of genotypes in the next generation will thus reflect the failures and successes of the preceding cohort[4] (i.e., what was "remembered" and what was "forgotten").

Darwin's striving

Since biological function is ultimately anchored in the evolutionary learning process which itself is the outcome of semiotically controlled interactions between millions and millions of individual organisms, it is not very surprising that biological functions always look "as if" they are teleologic in nature, for that is exactly what they are. It is therefore also no wonder that teleological language is popping up everywhere in biological literature even in scholarly work. When confronted with this embarrassing but broadly accepted use of language biologists or philosophers of biology usually respond by claiming that these ways of expression should be understood as only metaphorical, and that in due time, they can and will be reduced into the chemical and physical interactions that produce the effects or functions so labeled. Thus at a Wikipedia website on "function (biology)" we find the following typical response: "teleological terminology is often used by biologists as a sort of 'short hand' way of describing function, even though they know it is technically incorrect. *Laypersons may not understand this distinction, however*" (my italics, J. H.).

---

[4]While this simple pattern may indeed catch the biological learning process quite well as long as we are dealing with early states of evolution or with semiotically unsophisticated species, it is the opinion of this author that it gravely underestimates the importance of individual semiotic activity in later stages of evolution, particularly when brained animals are involved (cf. Hoffmeyer 2008).

The irony here is that Charles Darwin didn't understand this subtle distinction either. In Chapter 3 of *On the Origin of Species*, where Darwin describes the Malthusian character of the struggle for existence, he explicitly writes that: "In looking at Nature, it is most necessary [...] never to forget that every single organic being may be said to be striving to the utmost to increase in numbers" (Darwin 1971 [1859]: 71). This was certainly not meant as a "short hand" way of expression. Darwin's masterpiece was concerned with the origin of species, not with the origin of life, and he remained agnostic when it came to the question of the origin of life. That all the creatures of this world were incessantly "striving" to find food, escape predators, find mating partners etc. was so obvious to Darwin that he simply took this fact as a point of departure for his whole analysis. And in the very last two sentences of *On the Origin of Species* he even expresses the view that God created the first life forms:

> Thus, from the war of nature, from famine and death, the most exalted object which we are capable of conceiving, namely, the production of the higher animals, directly follows. There is grandeur in this view of life, with its several powers, having been originally breathed by the Creator into a few forms or into one; and that, whilst this planet has gone cycling on according to the fixed law of gravity, from so simple a beginning endless forms most beautiful and most wonderful have been, and are being evolved. (Darwin 1971[1859]: 463)

Among the "several powers", breathed into the first organisms by the Creator, Darwin of course counted "striving", or what we today would call *agency*.

Whether or not Darwin also in private believed in the Creator is not essential in this context. What is essential is that he had no trouble in accepting the idea that living beings possess agency and that his conception of the human situation, vis-à-vis nature, was in fact not very far from the biosemiotic one.

In the neo-Darwinist reinterpretation of Darwin's work which took place during the first half of the 20th century, the allusion to the Creator had to be abolished and with it disappeared Darwin's conception of "life, with its several powers". The task at hand here was to make a synthesis between the different branches of evolutionary biology inside a modern probabilistic understanding. Changes in a population now became seen as a movement through an "adaptive landscape" toward a state of equilibrium, where gene frequencies were stabilized. The individual organism lost its role as the leading evolutionary agent and became a slave to internal and external events and forces: the selective pressures from without and the mutational processes inside germ cells.

With the fusion of Darwin's theory with molecular genetics, a further burden was placed upon the shoulders of *natural selection*, a burden that slowly annihilated the original vision of Darwin. For now even thoughts and feelings had to be understood as genetically-based components in the behavior of an animal — involuntary reflexes not much different from the mechanical movements of a compass needle. Gone was the possibility of visualizing animals as *agents* in their own life.

So, while Darwin had placed the human being safely and understandably within the masterpiece of nature, the neo-Darwinist threw us out again — telling us that our *genes* may well belong to the reality of nature, but our *thoughts* and *feelings* are well outside the reach of science. Consequently, those previously least deniable aspects of our own biological experience — because they could not be reduced to a molecular explanation — came to be viewed with increasing suspicion, as if they were *Fata Morganas*, or epiphenomena without proper autonomous ontological reality (Hoffmeyer 2008).

Genes are segments of molecules, namely those long threads of nucleic acid bases coiled into a double helix structure which we call DNA. And to claim that molecules or segments of molecules possess agency is absurd. So, on the surface of it, the neo-Darwinian synthesis has indeed managed to get rid of the teleological aspect of life or to redefine it as survival strategies to be explained through game theoretical analysis. Properties of "winner genes" would prevail and "loser genes" would be lost.

The homunculus trap

That a majority of scientists seems satisfied with this hiding away of life's agential aspect behind a statistical curtain of competing "gene strategies" never stops surprising me: how can a segment of a molecule possibly have a strategy?! The majority is so overwhelming in fact that the most obvious conclusion would seem to be that I must be wrong.[5] While I will leave the judgment of this possibility with the reader, I must dare to continue my argument comforting myself, at least, with the fact that giants like Darwin himself and the English philosopher Bertrand Russell were on my side. Russell stated the matter very clearly when in 1960 he said that "every living thing is a sort of imperialist, seeking to transform as much as possible of its environment into itself and its seed. [...] We may regard the whole of evolution as flowing from this 'chemical imperialism' of living matter" (Russell 1960). Theoretical ecologist Robert Ulanowicz, from whom I have taken this quote, makes the further comment that "Russell, unfortunately, did not offer to explain the etiology of such 'imperialism', but it was obvious in the way Russell phrased his comment that he was *not* referring to the disembodied, external action of 'natural selection'" (Ulanowicz 2009).

---

[5]One of the few modern biologists to take note of this "homunculus gap", as John Deely pointed out in some detail in his 1969 discussion of species (Part i, pp. 105–115; Part ii, p. 293 and *passim*), was C. H. Waddington (e.g., 1960; 1961), with his insistence that the epigenetic system (i.e., the individual development and agency of the organism), and not only natural selection, acts as an anti-chance factor in evolution.

The chemical imperialism Russell talks about here is the same property that I have chosen to call agency, the capacity of a unit system to generate end-directed behaviors. The central point is that agency cannot be explained away by natural selection. This is because natural selection could not — as Darwin saw — work in the absence of "striving" or agency. Non-human organisms do not ever try to survive (for the simple reason, that they do not know that they shall die), but all of them do exhibit agency: they ceaselessly strive to find nourishment, find shelter, escape predators, find mating partners or whatever is necessary for them to do in their life. Without this agency fertility and competition would be pointless, and since the core of natural selection is the so-called "surplus-production" of individuals, i.e., the production of more individuals than can possibly survive under the given ecological conditions, natural selection would never get to do its work if the individual organisms did not exhibit agency.

The claim that the neo-Darwinian scheme explains function therefore rests upon a tacit smuggling in of a "homunculus", i.e., some unspoken agency hidden in the deep structure of the theory. Homunculus means "little man" and the concept was originally used to refer to attempts by alchemists to produce human beings in the laboratory through obscure experimental procedures, but the term came to fame when the first microscopists discovered small "animalcules" in the semen of humans and other animals. The idea then appeared that a little man, a homunculus, lay preformed inside the head of the sperm cell, and that this would explain human ontogeny: upon fusion with the egg cell the little homunculus just had to grow to adulthood (Fig. 2).

Homunculus theories are surprisingly widespread and were for instance common in early cognitive science where people were led to search for a "central processor" in the brain that was then somehow supposed to be responsible for human thinking. I remember learning at school that due to refraction a newborn baby would see the world upside down — until she learned to turn it around. This of

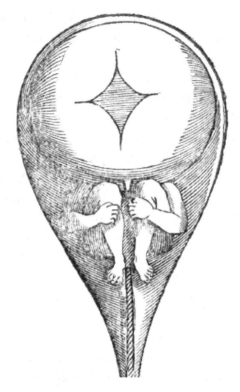

Fig. 2. Homunculus in sperm as drawn by Nicolaas Hartsoeker in 1695.

course presupposes that vision depends on somebody, a homunculus, sitting inside the brain looking at a screen showing the projected image.

The general problem with homunculi as theoretical entities is that they themselves need to be explained. The homunculi inside a sperm head would die childless unless its miniscule testicles would contain preformed sperm cells containing even smaller homunculi inside them, leading us to an infinite regress. The same goes for the homunculus in a baby's brain that in order to look at the inverted picture on the little brain-screen will have to have yet another small homunculus inside its own brain executing the same surprising activity.

Genes and semiotic scaffolding

We can see then that whenever biology uses semiotic terms such as information, adaptation, signal, messenger, fidelity, cross-talk, cue etc., it tacitly presupposes some homunculus-like principle to substitute for Darwin's "striving" (Kull *et al.* 2009). Essentially this is because biological function and semiosis are intertwined or even perhaps co-extensive. The evolutionary roots of agency, function and semiosis are buried in the now extinct prebiotic systems that approximately four billion years ago gave rise to the first life forms. Realistic modeling of such systems may give us indications of the concrete steps such a process may have taken, but for the moment we simply prefer to push back the origin question to a "threshold zone" below which we do not find semiosis, function or agency, and above which these properties are indeed exhibited by the system. Important work is presently done in this area by several groups (Kauffman, Clayton 2005; Deacon 2006).

Here we shall elaborate a little further on how semiosis and genetic fixation are intertwined in the learning process called natural selection (as if nature could select). I shall claim that this process becomes fully understandable only when the semiotics of the process is integrated into the analysis. The evolution of the hemoglobin family of proteins will illustrate this point.

In addition to the operational genes that most people equate with the very concept of *genes*, the genome contains huge stores of hidden, never sought-after "menu items" — what scientists call *pseudogenes* — that potentially may be brought into circulation, and thus may come to play a role in evolutionary changes. Pseudogenes are areas of the genome that are *nonfunctional* in the sense that they are not available to the transcription process — but which nevertheless exhibit remarkable similarity to known *functional* genes in their base sequences.

Before the mid-1970s, DNA was considered a very static molecule. But it has become very clear since then that there exist quite a number of mechanisms whereby sequences on the DNA molecule may be duplicated and even change positions on the chromosome. Such mechanisms are supposed to have led to the formation of *gene families* — i.e., families of narrowly related functional genes, as well as *pseudogenes*.

Pseudogenes are of evolutionary interest because, unlike functional genes, they are not expressed and, therefore, they are not subject to natural selection. Since they do not in any way contribute to the success or failure of their carrier organisms, they are free to undergo mutation without incurring the selectional consequences that normal genes do. (Normal genes mutate just as frequently, but those mutants that express functional deficiencies are accordingly sorted out.) Pseudogenes therefore represent a tacit *resource base* of latent proteins with unexplored properties — from which natural selection may eventually pick up new functional genes.

Human hemoglobin is in reality a complex (a *tetramer*) of four subunits, each of which is a protein chain of just under 150 amino acids. All four subunits are derived form the same ancestral protein chain but pair-wise they are now slightly different and are called *alpha* and *beta chains* respectively.

Significantly, the early embryo does not form any of these two subunits, but instead forms *xi* and *epsilon chains* that deviate somewhat from the alpha and beta chains of the adult hemoglobin. Towards the end of embryonic development, a fifth chain is made called the *gamma chain* — this is because the oxygen supply for the embryo comes through the placenta, and it therefore needs a special hemoglobin tuned to this temporary — but life-sustaining for that time period — condition. And for a short period around birth, the newborn baby even produces a sixth chain of proteins called the *delta chain*.

Such variety in the kinds of hemoglobin chain formation found through the embryogenesis of one individual is a relatively recent phenomenon, in terms of evolution — as is disclosed by comparing hemoglobin chains from a number of different species, as in Fig. 3.

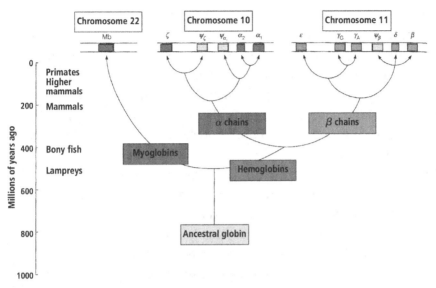

Fig. 3. An evolutionary tree for human hemoglobin. Above is shown the arrangement of human hemoglobin genes over three different chromosomes. Sequences that are still transcribed are shown in dark gray. The very light areas are no longer transcribed in humans, but are now pseudogenes. Below is shown a time axis (in millions of years) indicating approximately when in evolution different hemoglobin genes appeared (modified from Mathews *et al.* 1999: 234).

This figure shows that the original oxygen-carrying protein was a monomer not very different from the protein, *myoglobin* — a protein that nowadays is used for storing oxygen in muscle cells. Approximately five hundred million years ago, the gene involved in the protein synthesis of *myoglobin* underwent a duplication, where one of the gene copies became the ancestor of the *myoglobin* genes found in all higher organisms today, while the other gene copy, in time, developed (probably as a *pseudogene*) to become a proper oxygen transporter — and *this* gene thus became the progenitor of all the world's hemoglobin genes today. Thus, all later species in this lineage have separate genes for myoglobin and hemoglobin.

In retrospect, it is easy to see this evolutionary development as a reflection of a new "size strategy" in evolution. In the small animals of early evolutionary time (e.g., the protozoans and flatworms), there was not much need for an oxygen transporter, since the distance to the surface of the animal would always be insignificant. But as animals began to increase in size some 500 million years ago, and as a result began to feed on smaller animals for their nourishment, oxygen transportation became a major challenge. Four hundred million years ago, a second gene duplication occurred through which the ancestral forms of *alpha* and *beta chains* were established. Since then, all organisms in the lineage have both *alpha* and *beta chains* in their hemoglobin. In natural history, this corresponds with the divergence into sharks and bony fishes, and the evolutionary line of the latter that led to the reptiles and eventually to the mammals.

Similarly, the gene duplications behind the divergence of the *specialized fetal hemoglobin chains* occurred some two hundred million years ago, which roughly corresponds with the appearance of placental mammals — where the particular oxygen-binding properties of these chains would be needed for the special tasks of ensuring an adequate supply of oxygen to the embryo. And thus the numbering of the chromosomes in the upper part of Fig. 3 refers to the positions of the different hemoglobin genes on the human chromosome as they are today. Like the musical motifs of a classical symphony that are interwoven in a variety of ways to form the experience of a unified whole, so we see how evolution little by little has managed to weave the hemoglobin themes into one another to form a functionally optimal unity. But contrary to the product that is the symphony, here there is no composer behind this evolution, and its dynamics must be explained by means of natural processes.

Summarizing this case we can say that the decisive cause of the birth of a new functional gene would be a lucky conjunction of two events: (1) an already existing nonfunctional gene might acquire a new meaning through integration into a functional (transcribed) part of the genome, and (2) the gene product (e.g., a protein species) would hit an unfilled gap in the semiotic needs of the cell or the embryo. In this way, a new gene becomes a scaffolding mechanism, supporting a new kind of interaction imbuing some kind of semiotic advantage upon its bearer. It would thus be proper to refer to this phenomenon as *semiogenic scaffolding* (Hoffmeyer 2008). For by entering the realm of digitality, the gene's new semiotic functionality becomes available not only to the cells of the organism carrying it, but also to future generations — and, if we allow for horizontal gene transfer, even to unrelated organisms. Digitality in the life sphere assures the sharing (objectifying) of functions (and, in the human case, ideas), and thereby also their conservation through time.

But this very function is itself dependent on the relative *inertness* of the genetic material and its very indirect and highly sophisticated way of interfering with the worldliness of cellular life. Genes, like human words, do not directly influence the world around them (which is why we do not believe in the power of spells), but have whatever causal efficacy that they have *only* when — and for so long as — some living system *interprets* them. And just as inert, intrinsically meaningless words serve to *support* human activity and communication, so do inert, intrinsically meaningless genes *support* cellular activity and communication.

Biosemiotics and the theory of evolution

Biosemiotics implies a major change in our conception of the role of the genetic material, the genome. Whereas the traditional view sees the developmental process and thus the phenotype as specified by the genetic setup of the organism, the genotype, biosemiotics makes us consider the overall process as one of interpretation rather than one of specification. The so-called genomic "instructions" are buried deeply inside the nuclei of individual cells that cannot for good reasons "know" which destiny or function in the adult organism they are "meant" to occupy (a liver cell, a nerve cell, a lymphocyte etc.), and the cellular apparatus is therefore utterly dependent on external "instructions" to tell it which genetic "instructions" to execute at each moment. External cues derived from neighboring cells or tissues, or from body glands, or "traveling cells" such as the cells of the immune system (lymphocytes), must be picked up by specific receptors located on or in the cell membrane and translated to intracellular processes (signal transduction) that interact with the protein complexes involved in transcription and translation of DNA as well as with the targeting of virgin proteins either to their final positions in the landscape of the cell or to excretion into extracellular space. At each step in this complex chain of linked processes contextual factors may interact significantly with the outcome. Obviously the gene does not therefore in any ordinary sense of the word *determine* its own "reading", and neither is it justified to talk about this process as a *specification*. The process is best described as an interpretation with the implication that epigenetic and environmental factors must be seen as autonomous resources for ontogeny and phenotypic activity.

The fact that ontogeny is nevertheless very rigidly controlled has been taken as evidence for a dominating conservative mechanism linking the genetic setup to the functional phenotype through well-buffered chains of causal processes. In biosemiotics this explanation is supplemented or replaced by the notion of *semiotic scaffolding*: processes at all levels of organismic function are controlled by semiotic interactions between components that incessantly adjust biochemical or physiological activity to changing situations. This network of semiotic controls establishes an enormously complex semiotic scaffolding for living systems. Semiotic scaffolding safeguards the optimal performance of organisms through semiotic interaction with cue elements which are characteristically present in dynamic situations. And as we saw, at the cellular level, semiotic scaffolding assures the proper integration of the digital coding system (the genome) into the myriad of analogical coding systems operative across the membranes of cells and cell organelles. The big advantage of this mechanism is that contrary to physically based interactions, semiotic interactions do not depend on any direct causal connection between the sign vehicle and the effect. Instead the two events are connected through the intervention of an interpretative response. The point is that in semiotic interactions the causal machinery of the receptive system is itself in charge of producing the behavior, and it thus only needs to acquire a sensitivity towards the sign as an inducing factor. The biochemical machinery underlying the response is not, therefore, restricted by any bonds deriving from the chemistry of the releasing sign.[6] When functionality first arises at a holistic level (e.g., the level of the tissue rather than the individual cells making up the tissue), any mechanism that will stabilize such functionality will be favored by selection, and in this context semiotic interaction patterns provide fast and versatile mechanisms for adaptations, mechanisms that depend on communication and "learning"[7] rather than on genetic preformation.

The emergence of this sophisticated kind of stabilizing agency poses important challenges to our conceptions of evolution. In many cases semiotic scaffolding might work as a possible forerunner for natural selection: due to its stabilizing capacity semiotic scaffolding implies that natural selection may take routes that have been semiotically prepared for it rather than being directed by the haphazard events of fortunate chance mutations. Thus, biosemiotics holds that while genetic fixation may ultimately be the most secure mechanism for the stabilization of new traits or behaviors, such traits and behaviors were probably often initially stabilized through semiotic loops and thus kept functional for an indefinite period of generations, allowing the time needed for the much slower mechanism

---

[6]Thus any volatile compound, pheromone, may be taken up and used by the evolutionary process as a vehicle for the sign process whereby a certain behavior is released in any given insect no matter what the exact chemistry of that pheromone is. This absence of a direct involvement of the sign in the biochemistry and physiology of behavior of the receptive organism removes the hard limitations on causal mechanisms that would otherwise have to be obeyed.

[7]By learning at this level we refer to cells' acquisition of internal structural markers that reflect past events and influence future activity.

of supportive genetic change to take place. Especially in the later states of evolution such semiotic preparation is likely to have had an important influence on the actual outcome of selective solutions.

The introduction of a semiotic understanding has significant implications for ethology. Rather than reducing organismic behavior to inherited holistic lumps of reflexes, so-called instincts, biosemiotics suggests new subtle kinds of learning *in situ* associated with the establishment of semiotic scaffolding. The eventual appearance in big-brained animals such as birds and mammals of an experiential world would also be less of a mystery in the biosemiotic perspective than it is normally seen to be (and certainly not the meaningless epiphenomenon that it is claimed to be by some philosophers). Since the sign itself is a pure relation (and cannot be identified with any of the three entities it brings together), biosemiotics necessarily accepts the reality of a causality that is directed (but not effectuated) by the influence of pure relations, which suggests the possibility that the experiential worlds might be analyzed as a realm of pure relations connecting environmental cues and sensomotoric brain modules to contextual patterns dependent on memory and intentions.

At the ecological level biosemiotics requires us to extend our concept of an ecological niche to embrace the *semiotic niche*, i.e., the totality of cues around the organism (or species) which the organism (or species) must necessarily be capable of interpreting wisely in order to survive and reproduce. One important aspect of the semiotic niche concerns the semiotic interactions between individuals from different species. Most notably among these, of course, stands predator–prey relations, but just as significant perhaps is the range of interactive behaviors connected to symbiotic relations. When symbiosis is seen as a semiotically controlled kind of interspecific interaction it becomes evident that symbioses encompass a lot of subtle interaction forms that are not normally seen as belonging in this category. An illustrative case is so-called *plant signaling*. Undamaged fava beans (*Vicia faba*), for instance, immediately started to attract aphid parasitoids (*Aphidius ervi*) after having been grown in a sterilized nutrient medium in which aphid-infected fava beans had previously grown. The damaged beans thus had managed to signal their predicament through the medium to the undamaged beans, which then immediately started to attract aphid *parasitoids*, although no aphids were, of course, available for parasitoids to find.

The difference between this kind of *semiotic mutualism*, involving a delicate balance of interactions between many species, and symbiosis proper is one of degree rather than one of kind, and biosemiotics takes semiotic mutualism to be not only widespread in nature but nearly ubiquitous. This implies that the relative *fitness* of changed morphological or behavioral traits become dependent on the whole system of existing semiotic relations that the species finds itself a part of and, accordingly, the firm organism-versus-environment borderline will be dissolved, and a new integrative level intermediate between the species and the ecosystem would have to be considered — i.e., the level of the *ecosemiotic interaction structure* (Hoffmeyer 2008).

Clearly, this possibility becomes most interesting in cases where individual experience and learning enters the interaction pattern, as will often be the case in mammals and birds. Such learning might on occasion even subsume the evolutionary process, as is the case in human culture. Conversely, one might wonder if a relatively autonomous ecosemiotic interaction structure is precisely what is needed for learning to evolve in the first place. In this way, eventual increases in semiotic freedom, i.e., the capacity of a system to derive useful information from complex signs, will be prone to feed back into the evolutionary process by strengthening the advantages of possessing semiotic freedom. It follows that the evolutionary dynamics would possess an in-built tendency to invent and establish species exhibiting more and more *semiotic freedom* in the sense that their behavior would be increasingly underdetermined by the constraints of natural lawfulness, and increasingly dependent on the *in situ* interpretative capacity of individual organisms.

Biology and biosemiotics

It is indeed possible to explain non-human living systems as if they were deprived of semiosis, agency and true interests. This is exactly what biology has been doing for more than a century and often in ingenious ways. In the course of the last few decades, however, it has become clear that the reductionist assumptions that lead scientists to pursue such an awkward approach are no longer supported by scientific evidence. A new understanding of the thermodynamics of irreversible systems and of the dynamics of complex systems has given us a picture of a non-deterministic world in which bottom-up processes are interacting with top-down processes in intricate ways (Laughlin 2005; Kauffman 2008). This is — or ought to be — a great relief to biology because it implies that not everything can be predicted and that complex systems like life systems may evolve to create higher-order structures with properties that may have causal influence on lower-level entitles. In other words, it implies a view of nature as creative.

The order of things was not predictable from the outset but has to be explained by theories of a true emergence. We have argued in this paper that the emergence of life systems implied the simultaneous emergence of agency and semiosis as integral parts of the dynamics of such systems. A biology that wants to explain evolution cannot therefore fulfill its goal unless it integrates semiosis into its theoretical repertoire. In the absence of this explanatory tool, biology will never get rid of the hidden homunculi that unavoidably pop up underneath its mechanistic surface structure. Biosemiotics is necessary in order to make explicit those manifold assumptions imported into biology by such unanalyzed teleological concepts as function, adaptation, information, code, signal, cue, etc., and to provide a theoretical grounding for these concepts.

References

Darwin, Charles 1971 [1859]. *On the Origin of Species by Means of Natural Selection, or the Preservation of Favoured Races in the Struggle for Life*. London: J.M. Dent & Sons.

Deacon, Terrence 2006. Emergence: The hole at the wheel's hub. In: Clayton, Philip; Davies, Paul (eds.), *The Re-Emergence of Emergence: The Emergentist Hypothesis from Science to Religion*. Boston: MIT Press, 111–150.

Deely, John 1969. The philosophical dimensions of the origin of species. *The Thomist* 33(1): 75–149; 33(2): 251–342.

Deely, John 2001. *Four Ages of Understanding: The First Postmodern Survey of Philosophy from Ancient Times to the Turn of the Twenty-first Century*. Toronto: Toronto University Press.

Gould, Stephen Jay; Vrba, Elizabeth S. 1982. Exaptation — a missing term in the science of form. *Paleobiology* 8(1): 4–15.

Hoffmeyer, Jesper 2008. *Biosemiotics: An Examination into the Signs of Life and the Life of Signs*. Scranton and London: University of Scranton Press.

Kauffman, Stuart 2008. *Reinventing the Sacred: A New View of Science, Reason, and Religion*. New York: Basic Books.

Kauffman, Stuart; Clayton, Philip 2005. On emergence, agency, and organization. *Biology and Philosophy* 21: 501–521.

Kull, Kalevi; Deacon, Terrence; Emmeche, Claus; Hoffmeyer, Jesper; Stjernfelt, Frederik 2009. Theses on biosemiotics: Prolegomena to a theoretical biology. *Biological Theory* 4(2): 167–173.

Laughlin, Robert 2005. *A Different Universe: Reinventing Physics from the Bottom Down*. New York: Basic Books.

Mathews, Christopher; Holde, Kensal E. van; Ahern, Kevin G. 1999. *Biochemistry*. 3rd edition. San Francisco: Addison Wesley.

Russell, Bertrand 1960. *An Outline of Philosophy*. Cleveland: Meridian Books.

Ulanowicz, Robert E. 2009. *A Third Window: Natural Life beyond Newton and Darwin*. West Conshohocken: Templeton Foundation Press.

Waddington, Conrad Hal 1960. Evolutionary adaptation. In: Tax, Sol (ed.), *Evolution After Darwin: Vol. 1, The Evolution of Life*. Chicago: University of Chicago Press, 381–402.

Waddington, Conrad Hal 1961. Ch. 9: The biological evolutionary system. In: Waddington, Conrad Hal, *The Ethical Animal*. New York: Atheneum, 84–100.

Chapter 4
Biosemiotic Research Questions

Kalevi Kull, Claus Emmeche and Donald Favareau

Summary. We examine the biosemiotic approach to the study of life processes by fashioning a series of questions that any worthwhile semiotic study of life should ask. These questions can be understood simultaneously as: (1) questions that distinguish a semiotic biology from a non-semiotic (i.e., physicalist) one; (2) questions that any student in biosemiotics should ask when doing a case study; and (3) currently unanswered questions of biosemiotics. In addition, some examples of previously undertaken biosemiotic case studies are examined so as to suggest a broad picture of how such a biosemiotic approach to biology might be done. We introduce the terms "Sebeok's thesis", and "Uexküllian question".

Introduction

"What are the fundamental questions in biology?" is itself a question that remains pertinent in the examination of any given lineage of biological study.[1] "What are the fundamental questions in biology, if biology is seen as biosemiotics?" — i.e., as the study of sign processes that are essentially life processes — is the question that we wish to examine in some detail here.

Paradigmatic differences between the major approaches to biological investigation include differences in the questions that the

---

[1]See, e.g., Hacking (1983), Keller (2002), Levins, Lewontin (2006); etc.

respective types of inquiries will ask. Assuming that the semiotic approach in biology is, in fact, paradigmatically different from the reductionist and biophysical approaches that have been prevalent in biology since at least the Modern Synthesis of the 1930s, it will be fruitful to distinguish what we take to be the major differences between a semiotic and (what we will call for the purposes of this article) "non-semiotic" formulations of biological research questions. The successful framing of the relevant kind of biosemiotic questions, we will argue, may allow us to understand some aspects of life that a non-semiotic biology does not even inquire into (and therefore would never be able to successfully explain or describe).

We must mention at the outset, however, that while the purposes of this article are to contrast a semiotic approach to biology with a non-semiotic one, we see these not as alternative, but as complementary, approaches instead. Only by using both can we reach a more complete understanding of the phenomena and processes of life. Thus, since the principles and methodologies of the non-semiotic approach can be assumed to be well known to contemporary life scientists, the aim of this text will be to describe the biosemiotic approach by considering a series of questions that we feel any worthwhile semiotic study of life should ask. These questions can be understood *simultaneously* as: (1) the questions that distinguish a semiotic biology from a non-semiotic (i.e., reductionist–physicalist) one; (2) the questions that any student in biosemiotics should ask when doing a case study; and (3) currently unanswered questions of biosemiotics.

Semiotic versus non-semiotic life science

Semiosis, or true sign activity, occurs via a process of self-organization. Taking place in self-organizing systems, sign processes appear as emergent processes (or second-order self-organizing processes — which means the "organizing self") of signification and interpretation that co-ordinate the biochemical self-organization of living systems. In this sense, such activity might truly be thought of as being at the heart of the ongoing and interactive organizing of physical constituents into biological agents, or "selves".[2] Accordingly, such processes are not only upwardly causal (emergent) in their physical effects, but are also the result of downwardly causal or informational (semiotic) constraints upon the activity of the system as a whole.[3] It is the addition of this latter mode of self-organization to the former that distinguishes biotic from abiotic self-organizing systems. For semiosis — or the ability to create and take part in meaning-generating processes — is one of the distinguishing marks of a system that is alive. Many biosemioticians (e.g., Hoffmeyer 1997; Emmeche 2002; Kull 2000b) would go so far as to assert that it is not just "one of many", but is, in fact, the central distinguishing mark of any truly "living" system. Repeatedly formulated by Thomas A. Sebeok (1996; 2001) over the course of many decades, the concept that *life and semiosis are coextensive* we officially christen here as **Sebeok's thesis** — and it is one of the basic positions held in contemporary biosemiotics.[4]

As this thesis is at the heart of the entire "biosemiotic project" as we understand it, a few of Sebeok's more memorable formulations of the thesis follow:

> The process of message exchanges, or semiosis, is an indispensable characteristic of all terrestrial life forms. It is this capacity for containing, replicating, and expressing messages, of extracting their signification, that, in fact, distinguishes them more from the nonliving. (Sebeok 1991: 22)

---

[2]This view is compatible with, and in some senses generalizes, the idea of Polanyi (1968) that the information in the DNA acts as a boundary condition for the physical processes in the cell.

[3]See, e.g., Andersen *et al.* (2000), Pattee (2007).

[4]See also Anderson *et al.* (1984), Hoffmeyer (1997); also, "Thesis 1" in Emmeche *et al.* (2002: 14).

> All, and only, living entities incorporate a species-specific model (umwelt) of their universe; signify; and communicate by [...] signs. (Sebeok 1996: 102)
>
> Because there can be no semiosis without interpretability — surely life's cardinal propensity — semiosis presupposes the axiomatic identity of the semiosphere with the biosphere. (Sebeok 2001: 68)
>
> The life sciences and the sign sciences thus mutually imply one another. (Sebeok 1994: 114)

Taking this last formulation, especially, to heart, the first of our "meta-questions" becomes: *"How, precisely, should a semiotic life science (or biological sign science) proceed in its investigations?"* Certainly, some of the organizing principles and methodologies of life sciences that are already in place — such as naturalistic observation, *in vivo* experiments, qualitative reproducibility of findings, qualitative methods of study, etc. — can and will be made part of the daily work of the semiotic life scientist. But as the non-semiotic life sciences *de facto* rule out the viability of investigating such fundamentally semiotic phenomena as meaning, interpretation, subjective experience and sign relations, they can and do offer us no guidance on how to go about investigating such semiotic phenomena scientifically. The task is left to us, then — and in the short discussion that follows, we lay out a few of what we consider to be the most basic questions that a science studying semiotic phenomena in living systems will have to concern itself with.

Questions about the biosemiotics of organisms' experience and distinctions in umwelten

One such major question — i.e., *"How does the world in which any individual organism finds itself appear to that organism?"* — has been often perceived as inaccessible to scientific investigation and has therefore been left unresolved by reductionist biology. But in our opinion, scientific knowledge of other species' phenomenological experience need not be seen as any more *a priori* inaccessible than any other of science's previously "unsolvable mysteries". Rather, we feel that one of the main reasons why the question of organisms' subjective experience has been perceived as an "unscientific" question to begin with is simply because of the fact that "the scientific method" has been prematurely codified (and perhaps has subsequently become petrified) in a too narrow and restricted sense, reflecting its origins in seventeenth and eighteenth century mechanistic reductionism.

For while it is trivially true that we do not have a direct access to other organisms' (or even other human beings') phenomenological experience, it is important to take note that scientific knowledge is always "indirect" in some sense.[5] Neither physical nor semiotic knowledge can ever be "direct" or "immediate" — which is why it has been necessary to develop the many methodological apparatuses that enable traditional, non-semiotic science to work at all (these include the development of the rules of mediation between scientists and the objects of their inquiry, such as: the rules governing how to design an experiment, how to take measurements, how to analyse data, and how to draw conclusions about the mechanisms of the world, along with the more obvious technological apparatuses necessary for mediating the relation of observer to observed). In an analogous way, semiotic life science (or "biosemiotics") will likewise have to develop the proper mediating apparatuses — of both technology and of interpretation — proper to the successful examination, understanding, and explanation of the objects of its inquiry, accordingly.

---

[5] Cf. G. Bunge (1887): "Das Wesen des Vitalismus besteht darin, dass wir den allein richtigen Weg der Erkenntnis einschlagen, dass wir ausgehen von dem Bekannten, von der Innenwelt, um das Unbekannte zu erklären, die Aussenwelt. Den umgekehrten und verkehrten Weg schlägt der Mechanismus ein — der nichts anderes ist als der Materialismus — er geht von dem Unbekannten aus, von der Aussenwelt, um das Bekannte zu erklären, die Innenwelt."

Indeed, Rothschild (1962) had already realized that one of the main distinctions between biosemiotics and biophysics would lie in the difference of their adopted methods. Therefore the question: *"What are the specific methods of biosemiotics?"* when applied to the investigation of species-specific experience, becomes: *"What are the methods that allow the study of organisms' worlds (the umwelten)?"* Developing a biosemiotic approach thus means developing both quantitative and qualitative methods for biological research and analysis (Kull 2007a; 2007b). Via these methods, genuine scientific knowledge regarding both the ubiquitously observed semiotic phenomena (i.e., the sign relations qua "sign" relations) of organisms — as well as knowledge regarding the physically described substrates upon which such relations must necessarily take place — can, ideally, be obtained. But again, this will necessitate the introduction into science of something that non-semiotic science has generally resisted: an investigation into the qualities (or qualia) of experience and into the organization of subjective states *per se*.

Thus, biosemiotics can be generally defined as the *study of qualitative diversity found in and by living systems*. And in order to study such diversity, we will need to develop certain new methods — specifically biosemiotic methods — of research. Seen thus, the major task of any biosemiotic "case study" will be to ascertain the particulars of that available set of qualitative diversity that literally makes "sense" (i.e., that makes a "difference") for the organism or biosystem under study. In other words, the first task must be to describe the *umwelt* (or the entire network of experiential sign processes) proper to that organism. Thus we arrive at what we will call biosemiotics' **Uexküllian question**: *"How is the experiential world (or umwelt) of an organism organized?"* Or: *"In what does the world experienced by an organism (umwelt) consist?"*

With Charles S. Peirce, many of us in biosemiotics would reply to the second question: "in sign relations". And in this way, the general formulation of the Uexküllian question can also be considerably specified, both by Peircean semeiotic (logic of sign relations) and by Uexküllian *Umweltlehre* (see, for example, Deely 1994; 2002; Emmeche 2002; Favareau 2007; Hoffmeyer 1996b; Kull 2003; among many others). From that foundation, we can then fruitfully ask: *What would be a good general model to which most sign processes in living systems conform?*

One good proposal for such a model is the functional cycle (*Funktionskreis*) as described by Jakob von Uexküll in 1928. This model effectively conjoins an animal's sensing of the world with its subsequent actions upon the world — actions whose consequences for the organism are then fed back into the system cybernetically (we mean it here in the semiotic, not in the mechanical sense) in a recursively knowledge-generating loop. This is a feed-forward loop, according to Robert Rosen (1991; 1999), engendering anticipating agency into the physiology of the organism.

We suggest, therefore, that the model of functional cycle can serve as a usable initial general "working model" of semiosis in general and, of umwelt in particular. And, indeed, it has often been applied as such in the biosemiotic "case studies" that we will be considering at the end of this article.

Once adopted as an initial "working model" for the investigation into the varieties of naturally occurring biosemiosis, it is next worthwhile to ask: *"In what ways is it possible to improve upon the initial Uexküllian model?"* Happily, for it is a sign of a young science progressing, recent years have seen the proposal of several more detailed models of semiosis, many but not all of which use as their point of departure Jakob von Uexküll's original model (for an excellent review of these, please see Krampen 1997). Particularly, and as has often been noted at biosemiotic conferences and informal gatherings, it will be important for the continual development of our field to let our research guide us in developing the necessary amendments and alternatives to the Uexküllian *Funktionskreis* model for the experientially distinctive modes of iconic, indexical, and symbolic semiosis.

Questions about the biosemiotics of biological function

Umwelten — the experiential worlds within which organisms live and must choose their actions — are qualitative by their very nature. Accordingly, we need to develop an approach that can investigate such qualitative changes and differences in detail. For, once the reality of "qualitative" difference as such is accepted, as it is in biosemiotics, we then are faced with the scientific problems of determining the specific what, where, why, how, and when such qualitative differences actualize within and between organisms.

A non-semiotic life science approach generally will not deign such qualitative phenomena susceptible to scientific study, and can therefore offer little guidance or insight in this regard. Rather, such approaches will often go to quite extraordinary lengths to provide explanations of qualitative phenomena as being quantitative phenomena "in disguise".

To wit: In behavioural ecology, the communicative behaviour of animals is considered to be fully explained when its function for "the increase of the individual's fitness" has been substantiated in terms of the theory of natural selection. Thus Krebs and Dawkins (1984) use the phrase "mind-reading" (e.g., "a bird may mind-read a sneaking cat") as a "nakedly metaphorical" euphemism for those "underlying statistical laws" that they believe will predict what an animal will do as a consequence of its observation of another animal's behaviour. The "mind reader" function, they argue, is a Darwinian biological endowment that allows an organism to "optimize its own behavioural choices in the light of the probable future responses of its victim. A dog with its teeth bared is statistically more likely to bite than a dog with its teeth covered. This being a fact, natural selection or learning will shape the behaviour of other dogs in such a way as to take advantage of future probabilities, for example by fleeing from rivals with bared teeth" (Krebs, Dawkins 1984: 387).

Such statistical analysis reveals a major scheme of thought in contemporary behavioural ecology (as well as in the "classical ethologist view") that there are two principal causes of behaviour: the environmental and the genetic — and that these causes correspond to the two main types of manifest behaviour, learned and innate. A major contribution of biosemiotics in improving upon this scheme would be the provision of scientific explanations that would also account for cognitive–semiotic mechanisms of "qualitative difference" processing and negotiation in animal umwelten — an explanation (or, more likely, a coherent set of explanations) that would turn the frequently observed phenomena of "mind-reading" from a deliberately empty metaphor into a full-blown explanatory scientific concept (Kull, Torop 2003).

Relatedly, another of our overarching biosemiotic questions has to be: *"What are the general biological functions that are made possible through the phenomenon of semiosis?"* Minimally, these would seem to include: recognition, action choice, memory, code relations, categorization, and communication. A particularly fascinating question, moreover, concerns the nature of intentionality as a general feature of semiosic processes.[6] Deacon and Sherman (2008) have argued that such intentionality may start from the ability of living systems to recognize an absence — i.e., that the recognition of absence is what makes intention ("toward-ness") possible.[7] This, then, turns our attention to the empirical question: *"What sets of relations must necessarily be in place in order for an 'absence' to be recognized?"*

No less general than intentionality is the ability, in living systems, to categorize. As Lakoff and Johnson (1999: 17) have noted, "every living being categorizes". Thus, yet another empirical biosemiotic question then becomes: *"What are the processes by which organisms 'categorize'?"*[8] And because the ability to categorize

---

[6]Cf. Short (1981), Searle (1993), Deely (2007), Hoffmeyer (1996a).

[7]On the relatedness between intentionality, biological needs, and the recognition of absence, see also Kull (2000a).

[8]The fundamental role of the processes of categorization by organisms can be demonstrated via the view of a species as a "communicative category"

presupposes the ability to make distinctions: *"How are distinctions made by organisms and in organisms?"* Again, the wealth of scientific questions that have been left unanswered — primarily because they have been left unasked — by the non-semiotic life science approach has truly left "an embarrassment of riches" for 21st century semiotic life sciences to investigate.

Doing so, however, will entail grappling with some long-neglected fundamental questions right at the very start. An example of such a fundamental question is this: *"How can anything (e.g., molecule x) that initially does not have a function, obtain a function?"* Emmeche (2002) has argued that "obtaining a function" refers to the resetting of internal relations that occurs when a molecule (or some other structure) is taken in and made an integral part of a functional cycle or semiosis.[9] Questions that then arise from this most general formulation will then include: *"What are the primary biological functions?"*; *"How do these biological functions relate to specifically semiotic functions?"* and *"How may one kind of function turn into the other?"* And perhaps all of these last three questions may fall under the larger question: *"What are the major modes of biosemiosis?"*

It is important to notice that the kind of semiotic incorporation of qualitative change discussed by Emmeche (2002) is not fully reducible to the mere quantitative change that is also introduced with the inclusion of some new element (a molecule, etc.) into the operations of the body structure. Most such bioprocesses attain their particularly biosemiotic status from the dynamic feed-forward loop that they establish between both the organism's set of internal relations and the set of existing environmental relations "external" to the organism/environment interface (e.g., its membrane, skin or other

---

that has been developed in studies examining the evolutionary role of various forms of recognition between organisms and species (e.g., the "recognition concept of species" as developed by Paterson (1993), Lambert and Spencer (1995).

[9] On the biosemiotic concept of function, see Emmeche (2002).

boundary condition).[10] The model of the functional cycle demonstrates well the necessary node of the triadic sign relation that often lies outside of an organism's body.[11] Acknowledging the necessity of this agent–environment interdependence for the successful setting up and negotiation of sign relations makes it particularly interesting to ask: *"How is a particular semiosic process (meaning both any particular given 'sign' as well as the several different types of signs) extended spatially and temporally?"* and *"Which (physiological, ecological, and communicational) processes and structures are involved in this?"*

Perhaps, then, we spoke too soon when we noted earlier that such biosemiotic investigation would occupy 21st century science. For, as the empirically examinable and scientifically testable questions multiply exponentially (once one takes the relatively simple first step of refusing to believe *a priori* that all semiotic phenomena must be fully reducible to non-semiotic phenomena) several more centuries worth of productive scientific inquiry appear on the horizon.

Accordingly, we realize that we have yet but scratched the surface in our examination into the "questions" raised by and for a biosemiotic life science. So in the interest of brevity, let us just use what little remaining space we have at our disposal here to sketch out a few of the more pressing "big picture" questions that "taking the biosemiotic turn" engenders.

Questions about the attributes and boundary levels of biosemiosis

The biosemiotic approach to investigating life processes can in many ways be characterized as a relational approach, because its primary area of inquiry regards biological creatures' sign relations

---

[10] See Palmer (2004).

[11] Similarly to the concept of the "extended mind" in cognitive approaches (Clark, Chalmers 1998).

and the sign-relational aspects of those creatures' worlds. It is too, then — and perhaps on the most obvious level — an inquiry into many of the communicative aspects of living organizations, both within and between organisms. A fundamental question in "biosemiotic ethology" then will concern: "*How should science theoretically characterize the communication between (and within) organisms?*" Relatedly: A principled method for capturing and accounting for the translatability between sign systems occurring both between different species, and between different types, of semiosis (including human–animal communication) will certainly prove to be a pressing need once biosemiotic research advances to that point.[12]

Thus, the questions of development, from the biosemiotic point of view, turn out to be a deeply interconnected set of questions about the ways of categorization and re-categorization, as the introduction of new elements and the change or removal of old ones effect both local and global changes in the system's network of meaning (i.e., the relations between "instruction" and "appropriate response"), as such interdependent elements obtain new meanings (and new functions), and result (and become subjected to) different types of emergent sign processes. Perhaps the overarching question for semiotic life science here is: "*How to analyze living structures and organic forms themselves as 'communicative' structures?*"

Framed thus, the questions of evolution can then be seen simply as an inquiry into the fixation of these developmental changes, making them both irreversible for the species and part of the now-existing substrate upon which subsequent developmental change must take place (Hoffmeyer, Kull 2003; Kull 2000b). Such questions, when biosemiotically unpacked and related in detail to the rich spectrum of debates in contemporary evolutionary theory — including Evo-Devo and Developmental Systems Theory — should yield rich rewards in our understanding of the deep relations between life and change — an understanding that we can but gesture towards here.

---

[12]See Kull, Torop (2003).

It is hoped that a more biosemiotically informed approach to evolution will also throw some light on the questions regarding **semiotic thresholds**, or the major qualitative differences between the types of semiosis available to different organisms. The concept of a "semiotic threshold" was introduced originally by Umberto Eco (1979), who used that term to speak about the boundary between the semiotic and non-semiotic world.[13] Later, Terrence Deacon (1997) used the term "symbolic threshold" to differentiate between what he saw as the human-specific culture of thirdness relations (or "symbolic reference"), that is language, and the manifold number of other (iconic and indexical) sign systems used by all species (including humans) to gain knowledge about the world and (in some cases) to communicate to one another. The pressing question for this research agenda would then be *to determine the indexical threshold, and to describe the iconic and symbolic thresholds in further detail.*

Tackling the question from a more "categorical" perspective, we may reformulate the question in its "biggest picture" sense as: *"What are the main types and levels of semiosis?"* Kull and others have argued that the three main types of semiosis are, somewhat in the Aristotelian sense: the vegetative, the animal, and the rational (or propositional and lingual).[14] Here, we would hypothesize that the indexical threshold concerns the difference between vegetative and animal semiosis, though the more detailed reasoning behind this hypothesis must wait for another time. Obviously, there is work for biologically informed philosophers in the coming age of biosemiotic inquiry, as well!

---

[13]Eco (1979: 6) writes: "By natural boundaries I mean principally those beyond which a semiotic approach cannot go; for there is non-semiotic territory since there are phenomena that cannot be taken as sign-functions." On multiple approaches to such a semiotic threshold, see Stjernfelt (2003; 2007).

[14]Uexküll (1986a; 1986b); Emmeche (1984; 2004); Kull (2009); cf. Clarke (2003).

And finally, an ultimate — and perhaps the most immediately pressing and "practical" problem to be addressed by biosemiotics (and maybe by science and by human culture in general) — is *the problem of (bio)semiotic balance*. Organic balance, the balance of life, by its very nature is a semiotic balance. This means that the problem of ecological balance may converge with the problems of the balance of cultures and the problems of human health; thus the protection of biodiversity and protection of cultural diversity turn out to be parts of the same general problem — the protection of diversity, or quality as such (cf. Keskpaik 2001; Petrilli, Ponzio 2005).

Biosemiotic case studies

Many of the above questions have already begun to be addressed to some extent in a series of biosemiotic case studies undertaken during the past decade. We will not give a detailed review of all such studies here, but will only provide a short list that we think indicates some of the more important lines of research wherein contemporary empirical biology is interpreted biosemiotically, or where the biosemiotic approach is used as a major guide in understanding basic meaning-making mechanisms and/or significant patterns of sign relations in biosystems across many scales of integration.

1. The biosemiotics of animal communication (e.g., Sebeok 1972; 1977; Lestel 1995; Martinelli 2007a). Such study includes research into the sign behaviour of particular species (e.g., Sebeok 1994; Pain 2005). Of particular note here is a semiotic analysis of the vocal signs used by vervet monkeys that seeks to classify them according to the fundamental kinds of signs, infer minimum brain organizational constraints for their interpretation, hypothesize possible neuro-anatomical substrates able to satisfy these constraints, and propose an experiment based on the interpretation of signs in the monkeys in order to test these hypotheses (Queiroz, Ribeiro 2002).

2. Biosemiotic processes in ecosystems are crucial to proper ecosystem function. S. N. Nielsen (2007: 99) writes: "Where would

ecosystems be without insects to pollinate flowers? Bees could hypothetically be flying around in a random manner — which indeed would most likely lead to the result that some flowers would be fertilized. But adding their ability to smell flowers, see them at distance, possibly remember a good spot and for sure to communicate it to the other workers of the beehive, would increase the probability for this. These semiotic processes are crucial not only to the beehive but also to the ecosystem as such." Similarly, the abiding concern with the deep interrelations between ecosystems and organisms that was initiated in biosemiotics by its "precursors" such as Bateson (1979; 2000) and Uexküll (1928) has been carried on today by such "eco-biosemioticians" as Peter Harries-Jones and Myrdene Anderson. Additionally, Krampen (2001) has described the functional differentiation in ecosystems as a semiosic relationship, and Farina *et al.* (2007) have investigated the relationships of umwelten within an ecological landscape. The studies on the semiotics of plant and animal mimicry by Maran (2007) are also extremely worthwhile contributions to this field.

3. The biosemiotics of the immune system (Sercarz *et al.* 1988). Of recent note here is a semiotic model of signalling pathways in the B-cells of the immune system (El-Hani *et al.* 2007). This study substantiates the notion that semiotic modelling is required in order for the referential aspect of signalling processes to be analyzed and explicated. Hoffmeyer's extended discussion of the immune system's role in the creation of a self–nonself distinction in organisms is also relevant in this regard (Hoffmeyer 1996b).

4. The biosemiotics of signal transduction. A crucial part of comprehending the regulation processes within eukaryotic cells is the attempt to understand how specificity is determined (e.g., by the categorical sensing of the Ca^{2+} code), how and why ubiquitous messengers convey specific information, how the cell avoids undesired "cross-talk", and the role of redundancy for systemic integration. Bruni (2007) is particularly instructive in addressing all of these important issues.

5. Neurosemiotic approaches to brain research and consciousness studies have been proposed by Deacon (1997), Favareau (2002), Neuman (2003), Roepstorff (2004) and Villa (2005); while a biosemiotically informed approach to cognitive robotics has been undertaken by Ziemke and Sharkey (2001), Sharkey (1999; 2002), and Emmeche (2001).

6. A biosemiotic model of the genetic information system of the cell. Reconceptualizing the standard models of molecular biology for conceptualizing protein synthesis (including transcription, RNA processing, translation, etc.), several studies (Queiroz et al. 2005; El-Hani et al. 2006; 2008) adopt a Peircean model to explain such a sign process; Barbieri (2003; 2007) adopts an "organic code" model to explain the same. In both cases, the goal is to move away from reliance on the sterile and unhelpful use of the term "information" as a placeholder or mere metaphor in explaining genetic processes, and to replace it with a fully useable and genuinely bio-semiotic definition of what "information" for a living system consists in.

7. A biosemiotic taxonomy of systematic, compositional, sign-dependent relations in the living realm has been begun by Barbieri (2003; 2007) in his delineation of "the organic codes" of genetics — e.g., sequence codes, signal transduction codes, splicing codes, etc. The recently published *Codes of Life* (2008) features 18 different authors contributing to the development of this taxonomy, and building on this work, too, is Faria's (2007) study of how a major evolutionary change — in this case, the appearance of vertebrates — requires the development and fixation of new codes for the immune system.

8. Using Krampen's (1997) semiosic matrix, Huber and Schmid-Tannwald (2007) have devised a compelling reinterpretation of the process of oocyte-to-embryo transition in development by showing in detail how and in which sense the zygote acts as a "situated interpreter" not only of the inert nucleotides of the genome, but also of its specific context-dependent and context-constructing umwelt.

9. Comparative studies of sensorimotor interactions and inner representations in vertebrates (turtles, canids) and invertebrates (jellyfish, earthworms), discussing as well qualitative aspects of sensation (such as pain), have been surveyed from a general biosemiotic perspective in a fascinating study by Stephen Pain (2007).

10. Studies of vegetative semiosis, particularly in plants (Krampen 1981; Kull 2000a; 2000b; Baluška *et al.* 2006; Barlow, Lück 2007) have demonstrated the existence of genuine sign processes in the vegetative realm of life.

11. The role of sign types at various biosemiotic levels for the emergence of the human linguistic animal, and in particular differences and similarities between human language and animal communication, have been suggestively studied and biosemiotically analyzed by Terrence Deacon (1997), Donald Favareau (2008) and Dario Martinelli (2007b). John Deely's seminal *What Distinguishes Human Understanding?* (2002), and more recently *Intentionality and Semiotics* (2007), should also be noted in this regard.

Conclusion

Having in this article only briefly gestured towards just some of the major questions facing biosemiotic inquiry at this stage of its development, we hope we have yet been able to demonstrate some of the fundamental differences entailed between taking a semiotic approach to life science and the approach taken by non-semiotic biology, where such questions are provided no conceptual space in which to be asked. That said, one should not ignore the fact that from within that very successful tradition of non-semiotic biology, there have recently (and, indeed, increasingly) emerged new trends in genetics and in cellular and molecular biology that aim at more systemic and holistic understandings than had previously been attempted in the atomistic–reductionist phase of their sciences.

The search for such "higher-order understandings" is expressed not only by the increasing prevalence of key terms such as systems biology (Ideker *et al.* 2001), and the big -*omics* or mapping projects (e.g., genomics, transcriptomics, proteomics, cellomics, etc.) — but also by the increasingly acknowledged necessity for integrating the many current findings across disciplines in order to model the complex dynamics of life processes such as regulation, morphogenesis, communication, cell death, cell differentiation, the epigenetic processes, and in general, the massively interconnected processes that are biocomplexity and biodiversity.

Biosemiotics as a nascent scientific paradigm-shift welcomes these trends towards a more organicist biology (cf. Gilbert, Sarkar 2000) but also criticizes a tendency among some of them to remain mired in a self-defeating metaphysics wherein only the dyadism of brute physical interactions, as established within the study of physics and chemistry and as interpreted atomistically, are taken as explanatorily exhaustive of what is "real". For in order to explain such undeniably present phenomena in the biological world as subjective sensation, feeling, anticipation, awareness, meaning communication, and "mindedness", what is needed is a wider diversity of scientifically grounded concepts dealing with emergent *qualitative* novelties at different levels of biological integration. But this is precisely, of course, what a non-semiotic approach to biology, by definition, rules out the possibility of ever developing.

It is precisely in this domain of the life sciences, we believe, that biosemiotics can contribute to a better understanding of the interdependently *relational* and *semiosic* nature of the living being, and of the contextual settings determining the interpretations of intrinsic signs processes in complex biosystems. In this sense, a more qualitative form of the "organicist" framework is achieved, integrating the rich findings of non-semiotic research within an expanded (or extended) biology that can and will undertake the inquiry into the underlying "science of signs". For only within such a new biology can phenomena such as organic "qualia" be comprehended, including the qualitative organic relations characteristic of the human species.

This is so because, in the modern epoch of scientific research, the human species has been considered as, on the one hand, a strangely unique creature in the sense of having special access to reality through language and science, and of being capable of making objective descriptions of the world — and, on the other hand, as simply one species among others within the long and continuous history of evolution. As we see it, biosemiotics is an approach to the life sciences that makes it possible to unify these two apparently contradictory images of human beings and their place in nature. Our articulation of some of the preliminary biosemiotic "questions" that we have presented here will, we hope, help aid in the development of a focused set of conceptual tools that may eventually allow us to model the realities (both semiotic and non-semiotic) of not only the human experience of living being, but also the sphere of objectively significant organismal relations as categorized by other species, as well.[15]

References

Andersen, Peter Bogh; Emmeche, Claus; Finnemann, Niels Ole; Christiansen, Peder Voetmann (eds.) 2000. *Downward Causation: Minds, Bodies and Matter*. Århus: Århus University Press.

Anderson, Myrdene; Deely, John; Krampen, Martin; Ransdell, Joseph; Sebeok, Thomas A.; Uexküll, Thure von 1984. A semiotic perspective on the sciences: Steps toward a new paradigm. *Semiotica* 52(1/2): 7–47.

Baluška, Frantisek; Mancuso, Stefano; Volkmann, Dieter (eds.) 2006. *Communication in Plants: Neuronal Aspects of Plant Life*. Berlin: Springer.

Barbieri, Marcello 2003. *The Organic Codes: An Introduction to Semantic Biology*. Cambridge: Cambridge University Press.

Barbieri, Marcello 2007. Is the cell a semiotic system? In: Barbieri, Marcello (ed.), *Introduction to Biosemiotics: The New Biological Synthesis*. Dordrecht: Springer, 179–207.

---

[15] *Acknowledgements*. We thank Terrence Deacon, Jesper Hoffmeyer, and Frederik Stjernfelt for productive conversations on the formulation of these biosemiotic questions.

Barbieri, Marcello (ed.) 2008. *The Codes of Life: The Rules of Macroevolution*. Dordrecht: Springer.

Barlow, Peter W.; Lück, Jacqueline 2007. Structuralism and semiosis: Highways for the symbolic representation of morphogenetic events in plants. In: Witzany, Günther (ed.), *Biosemiotics in Transdisciplinary Contexts*. Vilnius: Umweb, 37–50.

Bateson, Gregory 1979. *Mind and Nature: A Necessary Unity*. New York: E. P. Dutton.

Bateson, Gregory 2000 [1972]. *Steps to an Ecology of Mind: Collected Essays in Anthropology, Psychiatry, Evolution, and Epistemology*. Chicago: University of Chicago Press.

Bruni, Luis E. 2007. Cellular semiotics and signal transduction. In: Barbieri, Marcello (ed.), *Introduction to Biosemiotics: The New Biological Synthesis*. Dordrecht: Springer, 365–408.

Bunge, Gustav 1887. Vitalismus und Mechanismus. In: Bunge, Gustav, *Lehrbuch der physiologischen und pathologischen Chemie*. Leipzig: F. C. W. Wogel, 3–15.

Clark, Andy; Chalmers, David J. 1998. The extended mind. *Analysis* 58: 10–23.

Clarke, David S. 2003. *Sign Levels: Language and its Evolutionary Antecedents*. Dordrecht: Kluwer.

Deacon, Terrence 1997. *The Symbolic Species*. London: Penguin.

Deacon, Terrence; Sherman, Jeremy 2008. The pattern which connects pleroma to creatura: The autocell bridge from physics to life. In: Hoffmeyer, Jesper (ed.), *A Legacy for Living Systems: Gregory Bateson as Precursor to Biosemiotics*. Dordrecht: Springer, 59–77.

Deely, John N. 2002. *What Distinguishes Human Understanding?* South Bend: St. Augustine's Press.

Deely, John N. 2007. *Intentionality and Semiotics: A Story of Mutual Fecundation*. Scranton: University of Scranton Press.

Eco, Umberto 1979. *A Theory of Semiotics*. Bloomington: Indiana University Press.

El-Hani, Charbel N.; Arnellos, Argyris; Queiroz, João 2007. Modeling a semiotic process in the immune system: Signal transduction in B-cells activation. *tripleC* 5(2): 24–36.

El-Hani, Charbel N.; Queiroz, João; Emmeche, Claus 2006. A semiotic analysis of the genetic information system. *Semiotica* 160(1/4): 1–68.

El-Hani, Charbel N.; Queiroz, João; Emmeche, Claus 2008. *Genes, Information, and Semiosis*. Tartu: Tartu University Press.

Emmeche, Claus 2001. Does a robot have an Umwelt? Reflections on the qualitative biosemiotics of Jakob von Uexküll. *Semiotica* 134(1/4): 653–693.

Emmeche, Claus 2002. The chicken and the Orphean egg: On the function of meaning and the meaning of function. *Sign Systems Studies* 30(1): 15–32.

Emmeche, Claus 2004. A-life, organism and body: The semiotics of emergent levels. In: Bedau, Mark; Husbands, Phil; Hutton, Tim; Kumar, Sanjev; Suzuki, Hideaki (eds.), *Workshop and Tutorial Proceedings. Ninth International Conference on the Simulation and Synthesis of Living Systems (Alife IX)*. Boston, 117–124.

Emmeche, Claus; Kull, Kalevi; Stjernfelt, Frederik 2002. *Reading Hoffmeyer, Rethinking Biology*. (Tartu Semiotics Library 3). Tartu: Tartu University Press.

Faria, Marcella 2007. RNA as code makers: A biosemiotic view of RNAi and cell immunity. In: Barbieri, Marcello (ed.), *Introduction to Biosemiotics: The New Biological Synthesis*. Dordrecht: Springer, 347–364.

Farina, Almo; Scozzafava, Silvia; Morri, Davide; Schipani, Ileana 2007. The eco-field: An interdisciplinary paradigm for ecological complexity. In: Witzany, Günther (ed.), *Biosemiotics in Transdisciplinary Contexts*. Vilnius: Umweb, 157–161.

Favareau, Donald 2002. Beyond self and other: On the neurosemiotic emergence of intersubjectivity. *Sign Systems Studies* 30(1): 57–100.

Favareau, Donald 2007. The evolutionary history of biosemiotics. In: Barbieri, Marcello (ed.), *Introduction to Biosemiotics: The New Biological Synthesis*. Dordrecht: Springer, 1–67.

Favareau, Donald 2008. Collapsing the wave function of meaning. In: Hoffmeyer, Jesper (ed.), *A Legacy for Living Systems: Gregory Bateson as Precursor to Biosemiotics*. Dordrecht: Springer, 169–212.

Gilbert, Scott F.; Sarkar, Sahotra 2000. Embracing complexity: Organicism for the 21st Century. *Developmental Dynamics* 219(1): 1–9.

Hacking, Ian 1983. *Representing and Intervening: Topics in the Philosophy of Natural Science*. Cambridge: Cambridge University Press.

Hoffmeyer, Jesper 1996a. Evolutionary intentionality. In: Pessa, Eliano; Montesanto, Anna; Penna, Maria Pietronilla (eds.), *The Third European Conference on Systems Science, Rome, 1–4 Oct. 1996*. Rome: Edizioni Kappa, 699–703.

Hoffmeyer, Jesper 1996b. *Signs of Meaning in the Universe*. Bloomington: Indiana University Press.

Hoffmeyer, Jesper 1997. Biosemiotics: Towards a new synthesis in biology. *European Journal for Semiotic Studies* 9(2): 355–376.

Hoffmeyer, Jesper; Kull, Kalevi 2003. Baldwin and biosemiotics: What intelligence is for. In: Weber, Bruce H.; Depew, David J. (eds.), *Evolution and Learning: The Baldwin Effect Reconsidered*. Cambridge: MIT Press, 253–272.

Huber, Johannes; Schmid-Tannwald, Ingolf 2007. A biosemiotic approach to epigenetics: Constructivist aspects of oocyte-to-embryo transition. In: Barbieri, Marcello (ed.), *Introduction to Biosemiotics: The New Biological Synthesis*. Dordrecht: Springer, 457–471.

Ideker, Trey; Galitski, Timothy; Hood, Leroy 2001. A new approach to decoding life: Systems biology. *Annual Review of Genomics and Human Genetics* 2: 343–372.

Keller, Evelyn Fox 2002. *Making Sense of Life: Explaining Biological Development with Models, Metaphors and Machines*. Cambridge: Harvard University Press.

Keskpaik, Riste 2001. Towards a semiotic definition of trash. *Sign Systems Studies* 29(1): 313–324.

Krampen, Martin 1981. Phytosemiotics. *Semiotica* 36(3/4): 187–209.

Krampen, Martin 1997. Models of semiosis. In: Posner, Roland; Robering, Klaus; Sebeok, Thomas A. (eds.), *Semiotics: A Handbook on the Sign-theoretic Foundations of Nature and Culture*, vol. 1. Berlin: Walter de Gruyter, 247–287.
Krampen, Martin 2001. No plant — no breath. *Semiotica* 134(1/4): 415–421.
Krebs, John R.; Dawkins, Richard 1984. Animal signals: Mind-reading and manipulation. In: Krebs, John R.; Davies, Nicholas B. (eds.), *Behavioural Ecology*. (2nd ed.) London: Blackwell, 380–402.
Kull, Kalevi 2000a. An introduction to phytosemiotics: Semiotic botany and vegetative sign systems. *Sign Systems Studies* 28: 326–350.
Kull, Kalevi 2000b. Organisms can be proud to have been their own designers. *Cybernetics and Human Knowing* 7(1): 45–55.
Kull, Kalevi 2003. Ladder, tree, web: The ages of biological understanding. *Sign Systems Studies* 31(2): 589–603.
Kull, Kalevi 2007a. Biosemiotics and biophysics: The fundamental approaches to the study of life. In: Barbieri, Marcello (ed.), *Introduction to Biosemiotics: The New Biological Synthesis*. Dordrecht: Springer, 167–177.
Kull, Kalevi 2007b. Life is many: On the methods of biosemiotics. In: Witzany, Günther (ed.), *Biosemiotics in Transdisciplinary Contexts*. Vilnius: Umweb, 193–202.
Kull, Kalevi 2009. Vegetative, animal, and cultural semiosis: The semiotic threshold zones. *Cognitive Semiotics* 4: 8–27.
Kull, Kalevi; Torop, Peeter 2003. Biotranslation: Translation between umwelten. In: Petrilli, Susan (ed.), *Translation Translation*. Amsterdam: Rodopi, 315–328.
Lakoff, George; Johnson, Mark 1999. *Philosophy in the Flesh: The Embodied Mind and its Challenge to Western Thought*. New York: Basic Books.
Lambert, David M.; Spencer, Hamish G. (eds.) 1995. *Speciation and the Recognition Concept: Theory and Application*. Baltimore: Johns Hopkins University Press.
Lestel, Dominique 1995. *Paroles de singes: L'impossible dialogue homme-primate*. Paris: Éditions la Découverte.
Levins, Richard; Lewontin, Richard 2006. *The Dialectical Biologist*. Cambridge: Harvard University Press.
Maran, Timo 2007. Semiotic interpretations of biological mimicry. *Semiotica* 167(1/4): 223–248.
Martinelli, Dario 2007a. Language and interspecific communication experiments. In: Barbieri, Marcello (ed.), *Introduction to Biosemiotics: The New Biological Synthesis*. Dordrecht: Springer, 473–518.
Martinelli, Dario 2007b. *Zoosemiotics: Proposals for a Handbook*. (Acta Semiotica Fennica 26.) Imatra: Finnish Network University of Semiotics.
Neuman, Yair 2003. *Processes and Boundaries of the Mind: Extending the Limit Line*. New York: Kluwer.
Nielsen, Søren Nors 2007. Towards an ecosystem semiotics: Some basic aspects for a new research programme. *Ecological Complexity* 4(3): 93–101.

Pain, Stephen P. 2005. The ant on the kitchen counter. *Journal of Biosemiotics* 1(2): 363–390.
Pain, Stephen P. 2007. Inner representations and signs in animals. In: Barbieri, Marcello (ed.), *Introduction to Biosemiotics: The New Biological Synthesis.* Dordrecht: Springer, 409–455.
Palmer, Daniel K. 2004. On the organism–environment distinction in psychology. *Behavior and Philosophy* 32: 317–347.
Paterson, Hugh E. H. 1993. *Evolution and the Recognition Concept of Species.* Baltimore: Johns Hopkins University Press.
Pattee, Howard H. 2007. The necessity of biosemiotics: Matter–symbol complementarity. In: Barbieri, Marcello (ed.), *Introduction to Biosemiotics: The New Biological Synthesis.* Dordrecht: Springer, 115–132.
Petrilli, Susan; Ponzio, Augusto 2005. *Semiotics Unbounded: Interpretive Routes Through the Open Network of Signs.* Toronto: University of Toronto Press.
Polanyi, Michael 1968. Life's irreducible structure. *Science* 160: 1308–1312.
Queiroz, João; Ribeiro, Sidarta 2002. The biological substrate of icons, indexes and symbols in animal communication: A neurosemiotic analysis of Vervet monkey alarm-calls. In: Shapiro, Michael (ed.), *The Peirce Seminar Papers: Essays in Semiotic Analysis*, vol. 5. New York: Berghahn Books, 69–78.
Queiroz, João; Emmeche, Claus; El-Hani, Charbel N. 2005. Information and semiosis in living systems: A semiotic approach. *SEED* 5(1): 60–90.
Roepstorff, Andreas 2004. Cellular neurosemiotics: Outline of an interpretive framework. In: Schult, Joachim (ed.), *Biosemiotik — praktische Anwendung und Konsequenzen für die Einzelwissenschaften.* (Studien zur Theorie der Biologie, 6.) Berlin: Verlag für Wissenschaft und Bildung, 133–154.
Rosen, Robert 1991. *Life Itself: A Comprehensive Inquiry into the Nature, Origin, and Fabrication of Life.* New York: Columbia University Press.
Rosen, Robert 1999. *Essays on Life Itself.* New York: Columbia University Press.
Rothschild, Friedrich S. 1962. Laws of symbolic mediation in the dynamics of self and personality. *Annals of New York Academy of Sciences* 96: 774–784.
Searle, John R. 1993. *Intentionality: An Essay in the Philosophy of Mind.* Cambridge: Cambridge University Press.
Sebeok, Thomas A. 1972. *Perspectives in Zoosemiotics.* (Janua Linguarum, Series Minor, 122.) The Hague: Mouton.
Sebeok, Thomas A. (ed.) 1977. *How Animals Communicate.* Bloomington: Indiana University Press.
Sebeok, Thomas A. 1991. *A Sign is Just a Sign.* Bloomington: Indiana University Press.
Sebeok, Thomas A. 1994. What do we know about semiosis in the domestic cat (*Felis catus*)? *Signifying Behavior* 1(1): 3–31.
Sebeok, Thomas A. 1996. Signs, bridges, origins. In: Trabant, Jürgen (ed.), *Origins of Language.* Budapest: Collegium Budapest 89–115.

Sebeok, Thomas A. 2001. Biosemiotics: Its roots, proliferation, and prospects. *Semiotica* 134(1/4): 61–78.
Sercarz, Eli E.; Celada, Franco; Mitchison, N. Avrion; Tada, Tomio (eds.) 1988. *The Semiotics of Cellular Communication in the Immune System*. Berlin: Springer.
Sharkey, Amanda J. C. 1999. *Combining Artificial Neural Nets: Ensemble and Modular Multi-Net Systems*. Berlin: Springer.
Sharkey, Noel E. 2002. Biologically inspired robotics. In: Arbib, Michael (ed.), *Handbook of Brain Theory and Neural Networks*. Cambridge: MIT Press.
Short, Thomas L. 1981. Semiosis and intentionality. *Transactions of Charles Sanders Peirce Society* 17(2): 197–223.
Stjernfelt, Frederik 2003. Sebeotics at the threshold: Reflections around a brief Sebeok introduction. *Semiotica* 147(1/4): 485–494.
Stjernfelt, Frederik 2007. *Diagrammatology: An Investigation on the Borderlines of Phenomenology, Ontology, and Semiotics*. (Synthese Library, 336.) Dordrecht: Springer.
Uexküll, Jakob von 1928 [1920]. *Theoretische Biologie*. 2te Aufl. Berlin: Springer.
Uexküll, Thure von 1984. Semiotics and the problem of the observer. *Semiotica* 48(3/4): 187–195.
Uexküll, Thure von 1986a. From index to icon, a semiotic attempt at interpreting Piaget's developmental theory. In: Bouissac, Paul; Herzfeld, Michael; Posner, Roland (eds.), *Iconicity. Essays on the Nature of Culture. Festschrift for Thomas A. Sebeok on his 65th Birthday*. Tübingen: Stauffenburg Verlag 119–140.
Uexküll, Thure von 1986b. Medicine and semiotics. *Semiotica* 61(3/4): 201–217.
Villa, Alessandro E. P. 2005. The neuro-heuristic paradigm. Paper Presented at the International Gatherings in Biosemiotics Conference 5, Urbino University, Italy. July 20–24.
Ziemke, Tom; Sharkey, Noel E. 2001. A stroll through the worlds of robots and men: Applying Jakob von Uexküll's theory of meaning to adaptive robots and artificial life. *Semiotica* 134(1/4): 701–746.

Chapter 5
Organism and Body: The Semiotics of Emergent Levels of Life

Claus Emmeche

Summary. This article comments upon some of the open problems in artificial life research from the perspective of a philosophy of biology tradition called qualitative organicism, and more specifically biosemiotics, the study of life processes as sign processes. Semiotics, in the sense of the pragmaticist philosopher and scientist Charles S. Peirce, is the general study of signs, and biosemiotics attempts to provide a new ground for understanding the nature of molecular information processing and sign processes at higher levels as well. Although we should not expect in Peirce to find any answers to the theoretical challenges and open questions posed by 'wet' artificial life, his semiotics (along with emergentist theories and cyborg studies) provide inspiration and conceptual tools to deal with the problems of life, mind, and information in the physical universe.

Introduction: Organicist philosophies

Artificial life research raises philosophical questions, just as cognitive science involves philosophy of mind. No clear demarcation line can be drawn between science and philosophy; every scientific research programme involves metaphysical assumptions and decisions about how to interpret the relations between experiment, observation, theoretical concepts and models (this was also evident when artificial life was originally formulated by C. G. Langton in the late 1980s; cf. Emmeche 1994). Yet we should not conflate questions that may be answered by science with questions that by their very nature are conceptual and metaphysical.

The aim of this paper is to address from the perspective of biosemiotics a subset of the open problems (as described by Bedau *et al.* 2000) raised by artificial life research, including "wet alife", about the general characteristics of life; the role and nature of information; how life and mind are related; and their relations again to culture and machines. Biosemiotics as the study of communication and information in living systems may provide some inspiration and conceptual tools for inquiry into such theoretical and philosophical issues. Firstly it is apt briefly to introduce organicism as a mainstream position in the philosophy of biology, and also a variant called qualitative organicism, and then introduce biosemiotics as a non-standard philosophy of biology.

Neither qualitative nor mainstream organicism are specific research paradigms; they are more like general and partly intuitive stances on how to understand living systems in the context of theoretical biology.

Organicism. In its mainstream form (cf. Emmeche 2001) organicism endorses these theses: (a) non-vitalism (no non-physical occult powers should be invoked to explain living phenomena); (b) non-mechanicism (living phenomena cannot be completely described merely by mechanical principles, whether classical or quantum); (c) emergentism (genuine new properties are characteristic of life as compared with "purely" physical non-living systems) implying ontological irreducibility of at least some processes of life (though methodological reductionism is fully legitimate); (d) the teleology of living phenomena (their purpose-like character) is real, but at least in principle explainable as resulting from the forces of blind variation and natural selection, plus eventually some additional "order for free" (physico-chemical self-organization). What is studied within an organicist perspective as emergent properties are seen as material structures and processes within several levels of living systems (developmental systems, evolution, genetic and biochemical networks, etc.), all of which are treated as objects with no intrinsic experiential properties. Mayr (1997) acknowledged his position as organicist, and mainstream organicism is widely accepted among biologists, even though the position was often mixed up with vitalism (see also El-Hani, Emmeche 2000; Gilbert, Sarkar 2000). Accordingly, there are no principled obstacles to the scientific construction of life and mind as emergent phenomena by evolutionary or bottom-up methods.

Qualitative organicism. This is a more radical position differing from mainstream organicism in its appraisal of teleology and phenomenal qualities. It emphasizes not only the ontological reality of biological higher level entities (such as self-reproducing organisms being parts of historical lineages) but also the existence of qualitative experiential aspects of cognitive behavior. When sensing light or colors, an organism is not merely performing a detection of external signals which then get processed internally (described in terms of neurochemistry or information processing); something more is to be told if we want the full story, namely about the organism's own experience of the light. This experience is seen as real. It may be said to have a subjective mode of existence, yet it is an objectively real phenomenon (Searle 1992 emphasized the ontological reality of subjective experience; yet, most of the time only in a human context). As a scientific stance qualitative organicism is concerned not only with the category of "primary" measurable qualities (like shape, magnitude, and number) but also with inquiry into the nature of "secondary" qualities like color, taste, sound, feeling, and the basic kinesthetic consciousness of animal movement. A seminal example of qualitative organicism is Sheets-Johnstone (1999). The teleology of living beings is seen as an irreducible and essential aspect of living movement, in contrast to mere physical change of position. This teleology is often attributed to a genuine form of causality ("final causation", cf. Vijver *et al.* 1998), and qualitative organicism's assessment of the "reality" of an instance of artificial life will partly depend on how to interpret the causality of the artificial living system.

Biosemiotics. The study of living systems from the point of view of semiotics, the theory of signs (and their production, transfer, and interpretation), mainly in the tradition of the philosopher and scientist Charles Sanders Peirce (1839–1914) but also inspired by the ethologist Jakob von Uexküll (1864–1944), has a long and partly neglected history in 20th century science (Kull 1999 for the history, Hoffmeyer 1996 for an introduction). It re-emerged in the 1990s and is establishing itself as a cross-disciplinary field attempting to offer alternatives to a gene-focused reductionist biology (much like one of the aims of artificial life, and indeed inspired by it), by gathering researchers for new approaches to biology, or a new philosophy of biology, ultimately with the hope to bridge the gap between science and the humanities. The semiotic approach means that cells and organisms are not seen primarily as complex assemblies of molecules, as far as these molecules — rightly described by chemistry and molecular biology — are sign vehicles for information and interpretation processes, briefly, sign action or *semiosis*.

A *sign* is anything that can stand for something (an *object*) in some interpreting system (e.g., a cell, an animal, a legal court), where "standing for" means "mediating a significant effect" (called the *interpretant*) upon that system. Thus, semiosis always involves an irreducibly triadic process between sign, object and interpretant. Just as in chemistry we see the world from the perspective of molecules, in semiotics (as a general logic of sign action) we see the world from the perspective of sign action, process, mediation, purposefulness, interpretation, and generality. Those are not reducible to a dyadic mode of mechanical action–reaction, or merely efficient causality. The form of causality governing triadic processes is final causation. Organisms are certainly composed of molecules, but these should be seen as sign vehicles having functional roles in mediating sign action across several levels of complexity, e.g., between single signs in the genotype, the environment, and the emerging phenotype.

Biosemiotics is a species of qualitative organicism for these reasons: (i) It holds a realist position regarding sign processes of living systems, i.e., signs and interpretation processes are not merely epistemological properties of a human observer but exist as well in nature, e.g., in the genetic information system (El-Hani *et al.* 2009). (ii) Biosemiotics interprets the teleology of sign action as related to final causation (Hulswit 2002). (iii) The qualitative and species-specific "subject" of an animal (i.e., its *umwelt* understood here as a dynamic "functional circle" of an internal representation system interactively cohering in action–perception cycles with an environmental niche) can to some extent be studied scientifically by the methods of cognitive ethology, neurobiology and experimental psychology, even though the experiential feeling of the animal is closed to the human umwelt (on umwelt research, see papers in Kull 2001). (iv) Signs have extrinsic, publicly observable as well as intrinsic phenomenal aspects. We can only access the meaning of a sign from its observable effects, a good pragmaticist principle indeed, but observation of the phenomenal experiences of another organism may either be impossible or highly mediated. However, reality exceeds what exists as actually observable. (v) Even though sign activity generally can be approached by formal and logic methods, sign action has a qualitative aspect as well. Due to the principle of inclusion (Liszka 1996) every sign of a higher category (such as a legisign, i.e., a sign of a type) includes a sign of a lower category (e.g., a sinsign, i.e., a sign of a token; a type has somehow to be instantiated by a token of it, just like any sign must be embodied). A symbol is not an index, but includes an indexical aspect, which again involves an icon. All signs must ultimately include (even though this might not be pertinent for their phenomenology) qualisigns, which are of the simplest possible sign category, hardly functioning as mediating any definite information, yet being signs of quality and thus having a phenomenal character of feeling (Peirce preferred a type–token–tone trichotomy for the type–token dichotomy; a tone is like a simple feeling). The argument (v) may strike a reader not acquainted with Peirce as obscure, but it is a logical implication of the ontological-phenomenological basis of Peirce's semiotics and points to an interesting continuity between matter, life and mind, or, to phrase it more precisely, between sign vehicles as material possibilities for life, sign action as actual information processing, and the experiential nature of any interpretant of a sign, i.e., the effects of the sign upon a wider mind-like system (Emmeche 2004).

To recapitulate, the biosemiotic notion of life is a notion of a complex web of sign and interpretation processes, typically with the single cell seen as the simplest possible autonomous semiotic system.

Synthetic biosemiosis? Computers are semiotic machines (Nöth 2003) and computers or any other adequate medium, such as a complex chemistry, can in principle function as a medium of genuine sign processes. Not all sign processes need to be biological, although all signs seem to involve at some point in their semiosis interpreters who typically would be biological organisms. Remember the distinction above between the interpretant as the effect or meaning of a sign and the interpreting system (or interpreter) as the wider system in which semiosis is taking place. So, what then is the biosemiotic stance regarding true synthetic life or "wet" artificial life? To answer this question, we have to consider, though more carefully than can be done here, (a) three non-exclusive notions of "life"; (b) the relation between the notions of organism, animal, body, and the general embodiment of various levels of signs processes; and (c) the semiotics of scientific models. This necessity of a precaution in assessment of the degree of genuineness of synthetic life in other media is related to another organicist theme: The thesis of irreducibility of levels of organization, or, as we shall interpret that thesis here, levels of embodied sign action.

"Life" in *Lebenswelt*, biology and ontology

Synthetic life provokes, of course, the general question "what is life?", partly because of an intuition we (or some people) have from our ordinary lives as some German philosophers would say, from the *Lebenswelt* (lifeworld) of human beings, that life (like death) is a basic condition we as humans can hardly control, know completely, or create. Now science seems to teach us otherwise. A contribution to clarify the issue is to be aware of the fact of the existence of at least three, non-exclusive notions of life. I will briefly sketch these:

***Lebenswelt* life.** A set of diverse, non-identical, culture-specific notions (determined by intuitive, practical, ideological, or social factors) of what it is to be "alive", what life and death is, why being alive and flexible is more fun than being dead and rigid, and so on. Science is distinct from, but not independent of, forms of the human *Lebenswelt* (just as scientific concepts can be seen as presupposing and being a refinement of ordinary language). For biological relevant notions of life, we can talk about "folk" and "experiential" biology (Emmeche 2000).

Biological life. The so-called life sciences are not interested in the life of the *Lebenswelt* as a normative phenomenon, but in the general physical, chemical and biological properties of life processes, as conceived of within separate paradigms of biology. This leads to several distinct "ontodefinitions" of life (Emmeche 1997; 1998)[1] such as life as evolutionary replicators, life as autopoiesis, or life as sign systems (and probably many more). However, advances in biotechnology and biomedicine will tend to mix up, "hybridize" or create new boundary objects (*sensu* Star, Griesemer 1989) between the domains of bioscience and a lifeworld deeply embedded in technoscience.

---

[1] See also the online version: "Defining life, explaining emergence", at http://www.nbi.dk/~emmeche/

Ontological life. Depending upon the ontology chosen, an ontological notion of life is marked by distinctions to other, similarly general and essentials domains of reality. Take the ontology of Peirce, for instance. Here there is a notion of life as instantiating the category of Firstness, it does not only include life in organisms, evolution or habit-taking (which are of the category of Thirdness); life is seen as an all-inclusive aspect of the developing cosmos, on par with spontaneity and feeling: "insofar as matter does exhibit spontaneous random activity (think of measurement error or Brownian motion), it still has an element of life left in it" (Reynolds 2002: 151).

Biosemiotics typically does not use Peirce's broad ontological notion of life, but construes a notion of life derived from contemporary biology, as mentioned, life as organic sign-interpreting systems. But biosemiotics entails a thesis of the reality of ideal objects, including possibilities like a fitness space, virtuality in nature, or tendencies in evolution and development, and so "the possibilities for final causes to prefer one tendency over another. Thus biosemiotics entails an ontological revolution admitting the indispensable role of ideality in this strict sense in the sciences" (Stjernfelt 2002: 342).

The invention of synthetic "wet" life may affect all of the three non-exclusive preoccupations with understanding life, that is, life within a cultural context, life as studied by science, and life as a metaphysical general aspect of reality. To approach how this may come about, we must analyze some levels of embodied life from the perspective of an emergentist ontology.

Level-specific forms of life and embodiment

As a species of organicism, biosemiotics is an emergentist position. However, it is not so often that emergent levels of sign processes have been explicitly discussed (Uexküll 1986; Kull 2000; El-Hani et al. 2009). The account given here should be seen as preliminary; the important point is not the number of levels (more fine-grained analyses may be done) but the very existence of separate levels of embodied sign action. See also Table 1.

The emerging forms of embodiment of life could be suspected merely to reflect a historically contingent division of sciences, an objection often raised against simple emergentist-level ontologies. Thus one should pay respect to the fallibilist principle and never preclude that new discoveries will fundamentally change the way we partition the levels of nature. The point is that from the best of our present knowledge we can construct some major modes of embodiment in which "life" and "sign action" plays crucially different roles, and in which we can place such broad phenomena as life, mind, and machines. Reflexivity is allowed for, so even the scientific description of these phenomena can be placed in this overall scheme of processes. A consequence is that wet artificial life is seen as a *hybrid* phenomenon of "the body of biology" and "the body of sociology", as will be explained below.

The emergent modes of embodiment, increasing in specificity (Table 1) are one-way inclusive and transcending: The human body *includes* the animate organism, which again presupposes multicellularity and basic biologic autopoiesis (but not vice versa). A human body (e.g., the body of a child, a soccer player, or a diplomat) as studied by anthropology is something more *specific* (i.e., in need of more determinations) than its being as an animal, thus transcending the mere set of animate properties (as having an umwelt) and organismic properties (like growth, metabolism, homeostasis, reproduction), just as an organism is a physical system, yet transcending the basic physics of that system. That an entity or process at an emergent level Z is *transcending* phenomena at level Y has two aspects. One is epistemic, i.e., "Z's description cannot adequately be given in terms of a theory generally accounting for Y, even though this Z-description in no way contradicts a description of the Y-aspects of Z." The other is ontological, i.e., "crucial properties and processes of Z are of a different category than the ones of Y, even though they may presuppose and depend on Y." Thus, a Z-entity is a highly

Table 1. Ordering relations between forms of embodiment. The epistemic dimension (top row) is shown by organizing those forms according to different domains of science each constituting its own objects; the ontic dimension (bottom row) is implied by an underlying ontology of levels of organization in nature. Increasing specificity from left to right; for each new level the previous one is presupposed.

The body of physics	The body of biology	The body of "evo-devo" research	The body of zoology	The body of anthropology	The body of sociology
Complex dissipative, self-organizing structures	Physiologic–homeostatic units with a genetic code-plurality, and irritability	Vegetative swarm of cells coordinating multicellular communication with multiple organic codes	Self-moving, action–perception cycles, animation, kinesthetics	Language, culture-specific *Lebenswelt*	The life in societal institutions, habitus formation

specific mode of realizing an Y-process, not explained by Y-theory. The organism is a physical processual entity with a form of movement so specific that physics (as a science) cannot completely account for that entity. The organism is a very special type of physical being, as it includes certain purposeful (functional) part–whole relations, based upon genuine sign systems of which the genetic code is the most well-known but not the only example. Here are brief characteristics of the levels.

"Life" as self-organization far from equilibrium

Physics deals with three kinds of objects: first, general forces in nature, particles, *general bodies* (matter in bulk), and the principles ("laws") governing their action; second, more specifically the structural dynamics of *self-organized bodies* (galaxies, planets, solid matter clusters, etc.); third, physical aspects of *machines* (artifacts produced by human societies and thus only fully explainable also by use of social sciences, like history of technology). One has often seen attempts to reduce all of physics to a formalism equivalent to some formal model of a machine, but there are strong arguments against the completeness of this programme (Rosen 1991), i.e., mechanical aspects of the physical world are only in some respects analogous to a machine. Some of the general properties of bodies studied in physics have a teleomatic character (a kind of directedness or finality, cf. Wicken 1987), which may be called "thermo-teleology", as this phenomenon of directedness is best known from the second law of thermodynamics (a directedness towards disorder), or from opposing self-organizing tendencies in far from equilibrium dissipative systems. Often when physicists talk about "life" in the universe the reference is to preconditions for biological life such as self-organizing non-equilibrium dissipative processes, rather than the following level.

Life as biofunctionality–organismic embodiment

A biological notion of *function* is not a part of physics, while it is crucial for all biology. Biofunctionality is not possible unless a living system is self-organizing in a very specific way, based upon a memory of how to make components of the system that meet the requirement of a functional (autopoietic and homeostatic) metabolism of high specificity. For Earthly creatures this principle is instantiated as a *code-plurality* between a "digital" genetic code of DNA, a dynamic regulatory code of RNA (and other factors as well), and a dynamic mode of metabolism involving molecular recognition networks of proteins and other components (see the semiotic analysis by El-Hani *et al.* 2009). Symbolic, indexical and iconic molecular sign processes are all involved in protein synthesis. The symbols (using DNA triplets as sign vehicles) seem to be a necessary kind of signs for a stable memory to pick out the right sequences for the right job of metabolism. This establishes a basic form of living embodiment, the single cell (a simple organism) in its ecological niche. This presupposes the workings of "the physical body" as a thermodynamic non-equilibrium system, but transcends that general form by its systematic symbolic memory of organism components and organism–environment relations.

Biosemiotics posits that organismic embodiment is the first genuine form of embodiment in which a system becomes an autonomous agent "acting on its own behalf" (cf. Kauffman 2000), i.e., taking action to secure access to available resources necessary for continued living. It is often overlooked that the subject–object structure of this active agent is mediated not only energetically by a structured entropy difference between organism and environment, but also semiotically, by signs of this difference; signs of food, signs of the niche, signs of where to be, what to eat, and how to trigger the right internal processes of production of organismic components at the right time. The active responsitivity of the agent organism (based upon observable molecular signs) has as an "inner" dimension, a quality of feeling, implied by what is in Table 1 called *irritability* at the level of a single cell. Irritability is a real phenomenon, well-known in biology, logically in accordance with a basic evolutionary matter–mind continuity, rationally conceivable, though impossible for humans to sense or perceive "from within" or empathetically know "what it feels like", say, for an amoeba or an *E.coli*.

It is highly conceivable that synthetic systems analogous to this level of embodiment may be produced some day.

Life as biobodies — coordinate your cells!

Characteristics like multiple code-plurality (involving the genetic code, signal transduction codes, and other organic codes, see Barbieri 2003) and forms of semiotic coordination between cell lines cooperating and competing for resources within a multicellular plant or fungus are characteristics of "the evolution of individuality" (Buss 1987). The "social" life of cells within a lineage of organisms with alternating life cycles constitutes a special level of embodied biosemiosis, and a special coupling of evolution and development. It is the emergence of the first *biobodies* in which the whole body constrains the growth and differentiation of its individual cells (a form of "downward causation", cf. El-Hani, Emmeche 2000). This multicellular level of embodiment corresponds to what was called a vegetative principle of life in Aristotelian biology, like that of a plant.

Life as animate — moving your self!

Here the body gets animated, we see a form of "nervous code" (still in the process of being decoded by neuroscience), and we see the emergence of animal needs and drives. When we consider animal mind and cognition, the intentionality of an animal presupposes the simpler forms of feelings and irritability we stipulate in single cells (including the "primitive" free-living animals, such as protozoa, lacking a nervous system), yet transcends these forms by the phenomenal qualities of the perceptual spaces that emerge in functional perception–action cycles as the animal's umwelt. Proprioceptive semiosis is crucial for phenomenal as well as functional properties of animation (Sheets-Johnstone 1999). More generally, the animal body is a highly complex and specific kind of a multicellular organism (a biobody) that builds upon the simpler systems of embodiment such as physiological and embryogenetic regulation of the growth of specific organ systems, including the nervous system. These regulatory systems are semiotic in nature, and rely on several levels of coded communications within the body and their dynamic interpretations (Hoffmeyer 1996; Barbieri 2003). The expression "the body of zoology" in Table 1 is used to emphasize both its distinctness as a level of embodiment, and that *zoology* instead of being simply part of an old-fashioned division of the sciences should be the study of animated movement, including its phenomenal qualities.

Life as anthropic — talk about life!

With the emergence of humans comes language, culture, division of work, desires (not simply needs, but culturally informed needs), power relations etc. The political animal not only lives and makes tools, but talks about it. Within this anthroposemiosis (Uexküll 1986), the body is marked by differences of gender (not simply sex), age, social groups, and cultures.

Life as societal — get a life!

After humans invented agriculture and states, more elaborated institutions could emerge and social groups became informed and enslaved by organizational principles of all the sub-systems of a civilized real society (work, privacy, politics, consumption, economy, law, politics, art, science, technology, etc.). Humans discover the culture-specificity of human life, "them" and "us". Reflexivity creeps in as civilization makes more and more ways to get a life. The body becomes societal (marked by civil life) and cyborgian (crucially dependent upon technology, machines). The political animal becomes cosmopolitical. The body is marked not only in the anthropic sense (see above), but also by institutions. *The cyborg body* is a civilized one, dependent upon technoscience (to keep "us" healthy and young) and, because of the dominant forms of civilization, ultimately co-determined by the globally inequal distribution of wealth. One can foresee artificial life research to play an increasing role in the contest over bodies and biopower as we approach the "posthuman" condition (cf. Kember 2003).

Hybridization and downward causation

This tour-de-force through some levels of embodiment makes a note on entanglement and hybridization relevant. The neat linearity implied by the concepts of inclusion and increasing specificity, and by the (admittedly idea-of-progress seeming) chain of levels does not hold true unrestricted. For instance, the very possibility of "human" creation bottom-up of new forms of life seems to suggest some complication (as human purposes may radically inform the natural teleology of what looks like biobodies). Already the culture-determined breeding of new races of cattle, crops, etc. suggests that even though biology should be enough to account for the body of a non-human animal, the human forms of signification interferes with pure biosemiosis, and create partly artificial forms of life like the industrialized pig or weird-looking pet dog races. In some deep sense, cows and pigs within industrialized agriculture are already cyborgs, partly machines, partly animals (cf. Haraway 1991). Culture mixes with nature in a "downward causation" manner, and thus, the hierarchy of levels is "tangled" (Hofstadter 1979), and "natural" and "cultural" bodies hybridize (Latour 1991). We might expect something similarly to apply if we access the status of "wet" artificial life, as reported by Rasmussen *et al.* (2004) and Szostak *et al.* (2001). Here, however, we need also to consider not only the biosemiotics of life, but also the special anthroposemiosis of experimental science, and especially the use of models and organisms to study life processes.

Models of life

Pattee (1989) was emphatic about the distinction between a model of life and a realization of some life process. In the early phase of artificial life research, focus was put on the possibility of "life in computers", and thus the question of computational simulations versus realizations was crucial. Considering the possibility of a "wet" bottom-up synthesis of other forms of life, we need to expand the kind of analysis given by Pattee to include not only the role of computational models in science in general and artificial life in particular, but also the very notion of a model in all its variety, and especially the notion of model organisms in biology. It is beyond the scope of this note to make any detailed analysis here, so in this final section only some hints will be given.

Let us make a preliminary, almost Borgesian classification of models in biology like the following.

Formal models and simulations. Highly relevant for "software" artificial life. Such models are, for their theoretically relevant features, computational and medium-independent, and thus disembodied, and would hardly qualify as candidates for "true" or "genuine" life, from the point of view of organicism or biosemiotics. Semiotically, the map is not the territory; a model is not the real beast.

Mechanical and ICT-models. The paradigmatic example here are robots. Robots may provide good clues to study different aspects of animate embodiment, but again, if taken not as models (which they obviously may serve as) but as proclaimed real "machine bodies" or "animats", their ontology is a delicate one. They are built by (often ready-made) pieces of information and communication technology;

they may realize a certain kind of "machine semiosis" (Nöth 2003), but their form of embodiment is radically different from real animals (see also Ziemke 2003; Ziemke, Sharkey 2001).

Evolutionary models. This label collects a large class of dynamic models not only across the previous two categories (because they may be either computational or mechanical, cf. also the field of "evolutionary robotics") but also combining evolutionary methods with real chemistries. Many sessions of previous artificial life conferences have been devoted to these models.

"Model organisms". The standard notion of a model here is to study a phenomenon, say regulation of cancer growth in humans, by investigating the same phenomenon in another but in some senses similar organism like the house mouse. In experimental biology, it has proved highly important to a fruitful research programme to choose "the right organism for the right job". *Drosophila* genetics is a well known case in point. The lineage or population of a model organism is often deeply changed during the process of adapting it to do its job properly, and it is apt to talk about a peculiar co-evolution of this population and the laboratories using it in research. For instance, Kohler (1994) describes how *Drosophila* was introduced and physically redesigned for use in genetic mapping and sees the lab as a special kind of ecological niche for a new artificial animal with a distinctive natural history.

"Stripping down" models. A method of investigating the minimal degree of complexity of a living cell by removing more and more genetic material to see how few genes are really needed to keep autopoiesis going (cf. Rasmussen *et al.* 2004). The problem, of course, as is well known from parasitology, is that the simpler the organism becomes, the more complex an environment is needed, so by adding more compounds to the environment, you can get along with fewer genes. The organism is always part of an organism–environment relation, which makes any single measure for complexity such as genome size problematic.

Bottom-up models. The term "bottom-up" may be used for all three major areas of artificial life, in relation to "software", "hardware" and "wet" models. Considering only the latter here (e.g., Szostac et al. 2001), the crucial question is to distinguish between, on the one hand, a process aimed at by the research that is truly bottom-up emergent, creating a new autonomous level of processes such as growth and self-reproduction pertaining to biofunctionality and biobodies; and on the other hand, something more similar to engineering a robot from pre-fabricated parts, that is, designing a functioning protocell but under such special conditions that one might question its exemplifying a genuine agent or organism. As exciting as they are as examples of advances in wet artificial life research, just as perplexing are they as possible candidates for synthetic true organisms, because their process of construction is highly designed by the research team. In this way, they are similar to the "model organisms" in classical experimental biology, but with the crucial difference that no one doubts the *latter* to be organisms, while it is question begging to proclaim the *former* to be.

The life-model entanglement problem

A special kind of hybridization is of interest here; the co-evolution of human researchers and a population of model organisms. As hinted at above, in the case of wet artificial life systems, "real" life and the model of life gets entangled. This raises questions not only about sorting out, or "purifying" as Latour (1991) would say, biosemiosis from anthroposemiosis to the extent that this is possible at all, but also considering in more detail the very nature of the entanglement. The hybridicity of human design "top-down" and nature's open-ended, evolutionary "design" bottom-up creates a set of complex phenomena that needs further critical study.

Conclusion

From an organicist perspective, real biological life involves complex part–whole relationships, not only regarding the structured network of organism–organs–cells relationships, but also regarding environment–(umwelt)–organism relations. The biosemiotic trend in organicism is needed to understand natural life (the plants and animals we already know) from a scientific perspective, but it is not enough to evaluate the complex question of "what is life?" as recently raised by synthetic chemistry approaches to wet artificial life. Here, more ontological, metaphysical, and philosophy of science (and scientific models) inspired considerations are also needed. Some of these have been presented here, others just hinted at.[2]

References

Barbieri, Marcello 2003. *The Organic Codes: An Introduction to Semantic Biology.* Cambridge: Cambridge University Press.

Bedau, Mark A.; McCaskill, John S.; Packard, Norman H.; Rasmussen, Steen; Adami, Chris; Green, David G.; Ikegami, Takashi; Kaneko, Kunihiko; Ray, Thomas S. 2000. Open problems in artificial life. *Artificial Life* 6: 363–376.

Buss, Leo W. 1987. *The Evolution of Individuality.* Princeton: Princeton University Press.

El-Hani, Charbel N.; Emmeche, Claus 2000. On some theoretical grounds for an organism-centered biology: Property emergence, supervenience, and downward causation. *Theory in Biosciences* 119(3/4): 234–275.

El-Hani, Charbel N.; Queiroz, João; Emmeche, Claus 2009. *Genes, Information, and Semiosis.* (Tartu Semiotics Library 8.) Tartu: Tartu University Press.

Emmeche, Claus 1994. *The Garden in the Machine: The Emerging Science of Artificial Life.* Princeton: Princeton University Press.

Emmeche, Claus 1997. Autopoietic systems, replicators, and the search for a meaningful biologic definition of life. *Ultimate Reality and Meaning* 20(4): 244–264.

---

[2] *Acknowledgements.* I thank Frederik Stjernfelt, Simo Køppe, Charbel Niño El-Hani, João Quieroz, Jesper Hoffmeyer, Mia Trolle Borup and Tom Ziemke for stimulating discussions.

Emmeche, Claus 1998. Defining life as a semiotic phenomenon. *Cybernetics and Human Knowing* 5(1): 3–17.
Emmeche, Claus 2000. Closure, function, emergence, semiosis and life: The same idea? Reflections on the concrete and the abstract in theoretical biology. [In: Chandler, Jerry L. R.; Vijver, Gertrudis Van de (eds.), *Closure: Emergent Organizations and Their Dynamics*.] *Annals of the New York Academy of Sciences* 901: 187–197.
Emmeche, Claus 2001. Does a robot have an Umwelt? Reflections on the qualitative biosemiotics of Jakob von Uexküll. *Semiotica* 134(1/4): 653–693.
Emmeche, Claus 2004. Causal processes, semiosis, and consciousness. In: Seibt, Johanna (ed.), *Process Theories: Crossdisciplinary Studies in Dynamic Categories*. Dordrecht: Kluwer, 313–336.
Gilbert, Scott F.; Sarkar, Sahotra 2000. Embracing complexity: Organicism for the 21st Century. *Developmental Dynamics* 219: 1–9.
Haraway, Donna 1991. *Simians, Cyborgs and Women: The Reinvention of Nature*. New York: Routledge.
Hoffmeyer, Jesper 1996. *Signs of Meaning in the Universe*. Bloomington: Indiana University Press.
Hofstadter, Douglas R. 1979. *Gödel, Escher, Bach: An Eternal Golden Braid*. London: The Harvester Press.
Hulswit, Menno 2002. From *Cause to Causation: A Peircean Perspective*. Dordrecht: Kluwer.
Kauffman, Stuart 2000. *Investigations*. Oxford: Oxford University Press.
Kember, Sarah 2003. *Cyberfeminism and Artificial Life*. London: Routledge.
Kohler, Robert E. 1994. *Lords of the Fly: Drosophila Genetics and the Experimental Life*. Chicago: The University of Chicago Press.
Kull, Kalevi 1999. Biosemiotics in the twentieth century: A view from biology. *Semiotica* 127(1/4): 385–414.
Kull, Kalevi 2000. An introduction to phytosemiotics: Semiotic botany and vegetative sign systems. *Sign Systems Studies* 28: 326–350.
Kull, Kalevi (ed.) 2001. *Jakob von Uexküll: A paradigm for biology and semiotics*. Berlin: Mouton de Gruyter (= *Semiotica* 127(1/4): 1–828).
Latour, Bruno 1991. *We have Never been Modern*. New York: Harvester Wheatsheaf.
Liszka, James J. 1996. *A General Introduction to the Semeiotic of Charles Sanders Peirce*. Bloomington: Indiana University Press.
Mayr, Ernst 1997. *This is Biology: The Science of the Living World*. Cambridge: Harvard University Press.
Nöth, Winfried 2003. Semiotic machines. *S.E.E.D. Journal (Semiotics, Evolution, Energy, and Development)* 3(3): 81–99.
Pattee, Howard H., 1989. Simulations, realizations, and theories of life. In: Langton, Christopher G. (ed.), *Artificial Life*. (Santa Fe Institute Studies in the Sciences of Complexity, 6.) Redwood City: Addison-Wesley, 63–77.

Rasmussen, Steen; Chen, Liaohai; Deamer, David; Krakauer, David C.; Packard, Norman H.; Stadler, Peter F.; Bedau, Mark A. 2004. Transitions from nonliving and living matter. *Science* 303: 963–965.

Reynolds, Andrew 2002. *Peirce's Scientific Metaphysics: The Philosophy of Chance, Law, and Evolution.* Nashville: Vanderbilt University Press.

Rosen, Robert 1991. *Life Itself: A Comprehensive Inquiry Into the Nature, Origin, and Fabrication of Life.* New York: Columbia University Press.

Searle, John 1992. *The Rediscovery of the Mind.* Cambridge: MIT Press.

Sheets-Johnstone, Maxine 1999. *The Primacy of Movement.* Amsterdam: John Benjamins.

Star, Susan Leigh; Griesemer, James R. 1989. Institutional ecology, 'translations', and Boundary Objects: Amateurs and professionals in Berkeley's Museum of Vertebrate Zoology, 1907-39. *Social Studies of Science* 19: 387–420.

Stjernfelt, Frederik 2002. Tractatus Hoffmeyerensis: Biosemiotics expressed in 22 basic hypothesis. *Sign Systems Studies* 30(1): 337–345.

Szostak, Jack W.; Bartel, David P.; Luisi, P. Luigi 2001. Synthesizing life. *Nature* 409: 387–390.

Uexküll, Thure von 1986. Medicine and semiotics. *Semiotica* 61(3/4): 201–217.

Vijver, Gertrudis Van de; Salthe, Stanley; Delpos, Manuela (eds.). 1998. *Evolutionary Systems: Biological and Epistemological Perspectives on Selection and Self-Organization.* Dordrecht: Kluwer.

Wicken, Jeffrey S. 1987. *Evolution, Thermodynamics, and Information: Extending the Darwinian Program.* Oxford: Oxford University Press.

Ziemke, Tom 2003. What's that thing called embodiment? In: Alterman, Richard; Kirsh, David (eds.), *Proceedings of the 25th Annual Meeting of the Cognitive Science Society.* Mahwah: Lawrence Erlbaum, 1305–1310.

Ziemke, Tom; Sharkey, Noel E. 2001. A stroll through the worlds of robots and men: Applying Jakob von Uexküll's theory of meaning to adaptive robots and artificial life. *Semiotica* 134(1/4): 701–746.

Chapter 6
Life Is Many, and Sign Is Essentially Plural: On the Methodology of Biosemiotics

Kalevi Kull

Summary. A comprehensive study of signs requires a consideration of ontology, paying attention to the kind and the way of existence of the objects under study. Biosemiotics, which attempts to develop a theoretical biology based on semiotics, would mean a redefinition of the concepts of general biology, taking into account the relational nature of life processes. A semiotic biology also means an application of semiotic methodology in empirical research. An aim of this chapter is to gather some ideas on characterization of (bio)semiotic methods of research. The distinction between a "physical eye" and a "semiotic eye" is characterized via the differences between the physical and semiotic reality, as two major complementary ways of scientifically built world-views. The difference between the physical and semiotic corresponds to the ontological difference between one and many, or a monist (quantitative, commensurable) and pluralist (qualitative, incommensurable) methodologies.

Biology means a study of living systems and life processes. As for any science, its task is to demonstrate the invisible in its domain (i.e., the aspects that are beyond everyday reality) — like the rules that order it, the components that build it, the forces that move it, the intentions that change it, etc. Sciences have developed many different tools or methods that are used for implementation of this task —

to uncover the invisible. Application of different toolboxes (like, for instance, by the humanitarian sciences or by the natural sciences) would mean different approaches, different ways of description. Biology can be different as dependent on the tools it uses. For instance, the biology that uses the tools or methods of physics is biophysics, and the biology that uses the tools or methods of semiotics is biosemiotics.

Phenomenal, physical, semiotic

In order to study living systems, it is required to identify the object — a living system, a life process. The identification may start from a pointing to anything that can be called "living systems" in an everyday (possibly rather professional) conversation or in common knowledge, in the reality that Humberto Maturana (1980) has called the "consensual domain". Science cannot avoid using this common phenomenal knowledge — it is part of communication between scientists, it is what we perceive; the scientist's orientation in his/her umwelt is based on this kind of phenomenal, everyday knowledge. However, science is not limited to it, and can study what is not directly visible. Telling of anything that is invisible means telling of anything that does not belong to the consensual domain or *everyday reality* (phenomenal reality), i.e., that which belongs to another domain or reality. These other realities can be, for instance, the *physical reality* (the universe), or the *semiotic realities* (the multiverses, in a certain sense). These can be called different realities because these are accessible *only* via the tools the scholarly approaches provide (theories, experiments, models, scientific translation and dialogue, etc.).

In the realm of physical reality, the concept of living system has to be independently defined — i.e., scientifically defined; one cannot just apply the concept of everyday reality. This is because the identification that uses a common practice of everyday language is insufficient in physics, due to the fundamental feature of the physical universe to exist independently of its interpretation. This is in accordance with the statement that everything (in the world according to the physical description — or, in physical reality) follows the universal physical laws, without exception.

Biophysical research has been able to delimit the area or distinctive features of living systems quite well. According to a biophysical definition, living systems are those that include a special type of autocatalytic processes — code-based reproduction. Code-based reproduction, however, is exactly the same process as interpretation, which means that life is an interpretation process.[1] This, as a matter of fact, would mean that life depends on its (own) interpretation which is a perfect argument for studying life not only in the terms of physical reality.

Biosemiotic studies demonstrate that living systems are those that make distinctions, or choose.[2] Since codes are the correspondence between different worlds,[3] and semiosis is what implements codes,[4] it has to be concluded that semiotic reality is multiverse, it includes many worlds — semiotic reality is many realities. Because of establishing and carrying on the code-based relations, the components of semiosis are "standing for something else" (according to Peircean understanding). A task of biosemiotics, accordingly, is to demonstrate the multitude, or meaningfulness, of the categories in living, and particularly, the invisible worlds of other organisms, the umwelten.[5]

---

[1] Cf. Chebanov (1993: 242), "life as interpretation process". Also, the concept of translation can be used in the same very general sense: "Translative processes pervade the entire living world, that is, the great biosphere" (Petrilli 2009: 553).

[2] Cf. Gregory Bateson's "difference that makes a difference". See also Hoffmeyer (2008).

[3] Cf. the definition of code by Marcello Barbieri (2001: 89): "a code can be defined as a set of rules that establish *a correspondence between two independent worlds.*" These are codes, in this sense, that make life irreducible, as formulated by Michael Polanyi (1968).

[4] On the principle of *code plurality*, see Emmeche (2004: 120); Kull (2007a: 173–174).

[5] Cf. Jakob von Uexküll's "niegeschaute Welten".

The principal feature of semiotic reality is the multitude or plurality of any object in it. This follows, almost trivially, from the nature of meaning — the meaningful object is not single, it is simultaneously anything else. Sign is an object that cannot be reduced to itself. Sign is always relational.

The sense of physical methods is to make a knowledge about physical reality possible. Physical reality is not what we see or feel, or recognize in our everyday behaviour. Physical reality is the quantitative universe that cannot be directly seen or touched by organisms. Only some organisms — the educated humans — can recognise physical reality (its invariance), doing this via experimental and theoretical methods of natural science, i.e., via constructing the physical (mathematical, quantitative) models of it, and using these models as representations of physical reality. Thus, only humans, via language ability, can represent physical reality. The discovery and description of physical reality is very much an achievement of the science of modernity.[6]

Analogically, the sense of semiotic methods is to make a knowledge about semiotic realities possible. Semiotic reality is not what we can entirely see or feel, or recognize in our everyday behaviour.[7] Semiotic reality is the qualitative multiverse that cannot be directly seen by organisms. Only some organisms — the educated humans — can recognise semiotic reality, (the multitude of any object), doing this via the methods of semiotics, i.e., via constructing the semiotic (logical, qualitative) models of it, and using these models as representations of semiotic reality. Thus, only humans, via the language-ability, can represent the semiotic reality.

The reality of being is different from both the physical and semiotic reality. This is the reality of action and perception, or umwelt, both personal and common, and this is available in some way for

---

[6]On the interpretation of the history of science from the point of view of semiotics, see Deely (2001), Favareau (2007).

[7]Cf. Deely (2009).

all organisms. It is given and does not require for its perception a scientific construction. However, in order to explain it, the complementary physical and semiotic realities have to be constructed.

Any example can be used to illustrate this distinction between the realities; for instance, a *border*. In the everyday reality of common sense, border is what we can point to as a border. In the physical reality, there are no borders *per se*; however, if to redefine the notion of border in a mathematical way, we can detect borders also in the physical reality. In semiotic reality, the semiotic border is the one that everybody is drawing differently but is identifying as the same; thus, any semiotic border is a multiple distinction that forms a category of the border.

The world as one is true for the physical reality. The reality of umwelten is the semiotic reality and is created in semioses. Since the interpretation process cannot be single, as communicative interaction always includes more than one code, the semiotic reality as the reality of umwelten is plural. Semiosis multiplies reality. In the strongest sense: semiosis can be defined as anything that multiplies reality.

Semiotic objects *versus* physical objects

Semiotic object is what by itself exists only due to possessing several meanings. The multiplicity of meanings is what makes it; a semiotic thing is the one that is simultaneously many things.[8]

Perfect examples of semiotic objects are provided by pictures that are designed to be ambiguous. For instance, as an illustration, let us use the well-known drawing "Message d'Amour des Dauphins" (1987) by Sandro Del Prete. For an adult human who sees the picture, it at first represents two naked people. For a child, it represents nine dolphins. For a dolphin, it probably represents neither one nor the other — maybe a bottle. For an educated human, it represents all

---

[8]Anything can be one, only if it has no meaning, i.e., if it is not a sign.

this and even more (including the "Message of Love", or just the material, or the ambiguity altogether, etc.).

However, what is *really* on that picture?

As for the physical reality, there is a certain spatial distribution of pigments on the surface. In this distribution of pigments, there is no any ambiguity. In a given moment of time, in a given light and temperature, there is just one distribution of molecules. There are no several patterns, there is just one heterogeneous molecular material. Physically, the "picture" is really *one*.[9] But this is not a picture, this is a pattern of matter.

As for the semiotic realities, there are many, both alternative and simultaneous, things on the picture. The number of potential objects on the picture is countless. Semiotically, the picture is really *many*. Once it is a picture, there are many pictures.

Semiotic objects cannot be static, because if they would be turned into anything static, they immediately lose the very nature of semioticity — their meaningfulness, their ambiguity, their turnaround in semiosis, their plurality.

An object is semiotic if it is in interpretation. Interpretation, according to the contemporary biosemiotic view, starts with the very process of life.

Physical objects are constructed by science as "dynamic structures" that are imagined as existing in a certain moment in one certain way, a single way.[10] A physical object may have several representations or models, but the object itself is thought to be existing as single.

Semiotic objects, in contrast do not exist in any single way. Their nature is existence in several ways ("at least two ways"[11])

---

[9]Until we start to measure it and to produce a sequence of descriptions — since the measurement process is a semiotic process.

[10]In the sense that would also include quantum physical objects.

[11]Referring with this expression to Juri Lotman's (1990) description of semiotic systems.

simultaneously.[12] While thinking on them as single we transform them into a physical object.

If we describe the variety of nature using a continuous scale of forms either in space or time or concerning the structural differences, we tend to approach the physical view on the world that usually uses monism. If we were to apply everywhere dual oppositions, we may threaten to be undistinguishable from dualism. But if we point to a series of thresholds and apply a more fine-grained classification of qualitative differences, we may find ourselves in the domain of semiotic approach, of pluralism.[13]

Likewise, a picture may have both physical and semiotic descriptions; any living system and life process can be approached in both ways.

The principal difference between biology and physics on the level of objects has been studied, e.g., by Walter Elsasser, Robert Rosen, Howard Pattee,[14] step by step coming closer to semiotics. Here, we develop the same path, now using a biosemiotic standpoint.

According to the biosemiotic understanding, the very nature of living systems lies in their ambiguity, in their plurality.[15] Cognitive objects can serve as a good example of semiotic objects.[16] These include, for instance, all perceptual categories, or *Merkzeichen* and *Wirkzeichen*, according to Jakob von Uexküll. But they also

---

[12] "Each thing is as many signs as there are grounds for distinct interpretations" (Short 1986: 106).

[13] Floyd Merrell wrote (in his letter to the author, October 28, 2006): "Most scholars do not believe Peirce was a pluralist, but I really don't think you take his theory of the sign as anything but pluralist: since the ultimate interpretant — its absolute truth — always escapes us, all our signs make up a pluralist set." See also Rosenthal (1994); Merrell (2007).

[14] See, for instance, Elsasser (1987); Rosen (1991); Pattee (1972).

[15] Cf. Sergey Chebanov's view on "the living being as a centaur-object" (Chebanov 1993: 225).

[16] See also Allen, Bekoff (1992).

include considerably more complex cognitive objects, like awakenness, dream, etc.

A very general class of semiotic objects can be called *categories*; perceptual categories as one of these.[17] Categories are the result of differentiation processes that occur due to interpretation, or semiosis. Good examples of categories as semiotic objects are the biological species as self-defining communicative systems.[18]

Semiotic modelling

A principal feature of semiotic objects concerns the hopelessness of deriving their differences and rules from the physico-chemical laws of nature, or from using general deductive models. Instead, their differences have to be discovered through a careful comparative analysis, an *ad hoc* classification. As, for instance, the alphabet or vocabulary of a language cannot be deduced from the physical laws of nature, the differences between words have to be found in a comparative study of uses of language.

The methodology of scientific description and analysis of physical systems is well developed. This is based on an assumption of the existence of universal laws of nature. These laws are non-contextual, i.e., they hold independently of the place, time, and situation. If there are natural laws that change in time or space, etc., then there exist more general non-contextual laws that describe these dependences in a precise way.

In the case of living systems, and in addition to the laws of nature, there appear regularities that are local and temporal and not deducible from the laws of nature. Examples of such regularities include habits, species, codes, and languages. These regularities (described as various rules and organic or cultural structures) are generally the products of communication. As different from the

---

[17]See also Stjernfelt (1992).

[18]On the semiotic concept of species, see, e.g., Kull (1992).

laws of nature and the structures deducible from these, the organic and cultural regularities are never universal, they include exceptions, they make errors, they change in time. However, of course, they never contradict natural laws — the latter simply do not determine the regularities in a predictable way; they instead provide the possibility of their existence in an immense variety.

What should be the result of a semiotic inquiry, and what is its difference from a physical description or a physical model?

First, a classification. Classical biology, like linguistics, provides many good examples of semiotic classification. Its methods owe much to Richard Owen's introduction of the types of homology and analogy as the basic tools of comparative research.

However, the resulting semiotic classification has to stay context-dependent. This means that any semiotic classification is a relative one, relational, and to a certain extent ambiguous. There cannot be absolute characteristics that describe certain species. A species is established in relation to other, similar species (externally), and by the individuals that recognize each other (internally).

Second, a semiotic description is a set of complementary descriptions. Niels Bohr has emphasized well the role of complementarity in description; however, he did not draw a clear distinction between the physical and semiotic reality, since the complementarity is necessary only for the latter. Physics may live without complementarity,[19] semiotics can not.

Third, a semiotic model has to be based on several mutually nonconvertable (noncommensurable) qualities, on qualitative differences. Thus the basis of semiotic models is not quantitative, it is qualitative.

Of course, we may find many semiotic features in the physical models and modelling. This is evidently so because science, including physics, has to deal with interpretation as a part of its work. However, in the case of interpretation of interpretation — and this

---

[19]Until it will not aim to describe the measurement process in full.

is what semiotics is and does — the fundamental features of the interpretation process are included in the object level itself, and this makes it different from the way physics looks at the world.

Different as the physical and semiotic realities are, it is possible to move between them in the scientific realm. Collapsing the multitude of an object into one thing,[20] or for a living organism to die, or to move from a qualitative description to a measurement, is the way to reach physical from semiotic. The opposite direction — from the physical to semiotic — would appear if describing everything as meaningful, or applying exclusively qualitative descriptions, or getting life from non-life. The relationship is obviously asymmetrical, the latter direction being more difficult than the former. However, it is certainly possible, also on the basis of a physicalist approach, to describe the conditions in which in some region of the universe an interpretation arises (which will at once be an interpretation of interpretation) and the plurality of an object will appear. These attempts have been made, for instance, by the theory of complexity. Thus, the semiotic can be seen as arising from a special case of physical, but also in reverse, the physical can be seen as a degenerated case of semiotic.[21]

Comments on semiotic methods

As Vyacheslav V. Ivanov pointed out as early as at the outset of the Tartu–Moscow school of semiotics, "The fundamental role of semiotic methods for all the related humanities may with confidence be compared with the significance of mathematics for the natural sciences" (Ivanov 1978 [1962]).

The methods of semiotic study or semiotic analysis have been described, more or less distinctively, not so often, and if they have,

---

[20]See L. Kauffman (2005), from whom this expression is taken.

[21]From a philosophical point of view, this position can be named as "local pluralism" (see also Kull 2009).

then in quite different ways by different semiotic schools (e.g., Clarke 2003; Maasik, Solomon 1994; Manning 1987).

Tommi Vehkavaara has emphasized that the explication of methods may be a central problem in biosemiotics nowadays (e.g., Vehkavaara 2002). Moreover, according to Vehkavaara, the study of methodological problems may turn out to be crucial in the development of biosemiotics.

Biosemiotics, in many ways, differs from the more classical fields in semiotics, like for instance semiotics of culture. A particular difference concerns their methods, since the study of sign processes in non-human organisms or inside the organism's body may not have much use in traditional methods of humanities or social sciences. Also, biosemiotic studies often appear in opposition to the alternative, natural scientific approaches used for studying the living systems.

Friedrich S. Rothschild, who was the first to use the term "biosemiotic" in 1962, emphasized the distinction between biosemiotics and biophysics on the basis of the differences in their methods (Rothschild 1962: 777):[22]

> We speak of biophysics and biochemistry whenever methods used in the chemistry and physics of lifeless matter are applied to material structures and processes created by life. In analogy we use the term biosemiotic. It means a theory and its methods which follows the model of the semiotic of language. It investigates the communication processes of life that convey meaning in analogy to language.

Robert Rosen's (1999) point of departure in his search for an adequate methodology for the study of living processes uses the opposition of this methodology to the physicalistic methods. Rosen describes *mimesis* as an alternative to reductionistic method.

---

[22] See also Rothschild (1968).

Thure von Uexküll, in his review of umwelt research as developed by Jakob von Uexküll, points to the specificity of its methods. He writes (T. v. Uexküll 1992: 281):

> *The approach of Umwelt-research*, which aims to reconstruct creative nature's "process of creating", can be described as "participatory observation", if the terms *participation* (*Teilnahme*) and *observation* (*Beobachtung*) are defined more clearly.

Observation can be characterized as follows (T. v. Uexküll 1992: 281):

> Observation means first of all ascertaining which of those signs registered by the observer in his own experiential world are also received by the living being under observation. This requires a careful analysis of the sensory organs (receptors) of the organism in question. After this is accomplished, it is possible to observe how the organism proceeds to decode the signs it has received.

Participation is defined (T. v. Uexküll 1992: 281):

> Participation, therefore, signifies the reconstruction of the *Umwelt* ("surrounding world") of another organism, or — after having ascertained the signs which the organism can receive as well as the codes it uses to interpret them — the sharing of the decoding processes which occur during its behavioral activities.

Acquiring knowledge of semiotic systems has to be based on the semiotic processes themselves — i.e., on communication, on dialogue. The knowledge a scientist can get in this way about the worlds of other organisms is of the type one can get via translation, or dialogue.[23]

---

[23] On the concept of translation as applied to interspecies communication, see Kull, Torop (2003).

Therefore, biosemiotics includes hermeneutics and has to use hermeneutic methods (Chebanov 1993; Markoš 2002). However, this kind of hermeneutics — *biohermeneutics* — cannot get much use out of Martin Heidegger's or Hans-Georg Gadamer's work — at least no more than some good hints, because these studies are strictly limited to the world of the languaging organisms, whereas what semiotic-hermeneutic biology has to be able to learn is to *translate* between the sign systems that are generally not languages.

Semiotics as a basically non-quantitative science would accordingly get much use in a wide and systematic application of qualitative methods (Anderson, Merrell 1991; Shank 1995; Kaushik, Sen 1990). The latter study includes participant observation, phenomenological description, individual case studies, structural analysis, naturalism, etc. However, a study on the application of qualitative methods as developed in social sciences is still almost absent in biosemiotics.

There have also been attempts to build what has been called a "qualitative physics" (Forbus 1988; Petitot, Smith 1990). Qualitative physics does not mean a claim that the physical reality is qualitative (a set of qualities), not at all. Quantitative physics has a solid fundament to pursue that the physical reality is commensurable and lawful. However, to convey this knowledge, the teaching of physics requires people's language to be involved, and that is the part where qualitative physics (or semiotics in physical teaching) is needed.

What specifies semiotic biology? What characterizes a biosemiotic inquiry?

A general purpose of semiotics is to study how choices are being made in living systems because semiosis can be seen as a general mechanism of choice.[24] Accordingly, a semiotic study may include two basic aspects.

---

[24] I use this definition in order to indicate that semiotics, indeed, provides a general understanding of the phenomenon of choice, and emphasize the radical difference from the determinism of physicalist models. The obvious

First, it describes structures, categories, habits, and codes that living systems have formed (or are forming) in their (inter- and intraorganismic) communicative processes.

Second, it models the interactions of habits, or the processes of behavioural choices.

The first means the application of structuralist, semiological methods, also nomothetic. The second means idiographic methods.

When using the term "scientific" here, I include under it both nomothetic and idiographic research. Therefore, in this sense, humanities are also sciences, or in German terms — *Geistwissenschaften* are also true *Wissenschaften*. These are *Wissenschaften*, but of a different type.

A *sine qua non* for a biosemiotic approach is the inclusion of the activity of organism, or subjectivity, as anything real and describable. The inclusion of subjectivity is done in terms of sign processes. Semiosis, a sign process, is itself the mechanism of subjectivity, which is the mechanism of making choices.

A single sign process is almost unreachable. However, a single text, a unique organism, a unique culture is approachable for a scientific study.[25] Here we are in the limits of idiographic science.

Thus, stating that living systems are multireal, or plurireal, it should be possible to take this principal nature of theirs into account. A physical approach describes everything in the world as unireal, including organisms. Description of organisms as plurireal would then mean (and assume) a semiotic description.

Biosemiotics as extended biology — the biology that can study the organism's subjectivity — would not exclude the methods of natural sciences, but it does not restrict itself there.

---

relationship between taking habits and making choices provides this possibility to replace one with the other in the definition.

[25]Cf. Kull (2002).

Thus, a semiotic approach means an enrichment of the toolbox that a biologist is allowed to use.[26]

References

Allen, Colin; Bekoff, Marc 1997. *Species of Mind: The Philosophy and Biology of Cognitive Ethology*. Cambridge: MIT Press.

Anderson, Myrdene; Merrell, Floyd 1991. Grounding figures and figuring grounds in semiotic modeling. In: Anderson, Myrdene; Merrell, Floyd (eds.), *On Semiotic Modeling*. (Approaches to Semiotics 97). Berlin: Mouton de Gruyter, 3–16.

Barbieri, Marcello 2001. *The Organic Codes: The Birth of Semantic Biology*. Ancona: Pequod.

Chebanov, Sergei V. 1993. Biology and humanitarian culture: The problem of interpretation in bio-hermeneutics and in the hermeneutics of biology. In: Kull, Kalevi; Tiivel, Toomas (eds.), *Lectures in Theoretical Biology: The Second Stage*. Tallinn: Estonian Academy of Sciences, 219–248.

Clarke, David S. 2003. *Sign Levels: Language and Its Evolutionary Antecedents*. Dordrecht: Kluwer Academic Publishers.

Deely, John 2001. *Four Ages of Understanding*. Toronto: University of Toronto Press.

Deely, John 2009. *Basics of Semiotics*. 5th edition. (Tartu Semiotics Library 4.2.). Tartu: Tartu University Press.

Elsasser, Walter M. 1987. *Reflections on a Theory of Organisms: Holism in Biology*. Baltimore: Johns Hopkins University Press.

Emmeche, Claus 2004. A-life, organism and body: The semiotics of emergent levels. In: Bedau, Mark; Husbands, Phil; Hutton, Tim; Kumar, Sanjev; Suzuki, Hideaki (eds.), *Workshop and Tutorial Proceedings. Ninth International Conference on the Simulation and Synthesis of Living Systems (Alife IX), Boston Massachusetts, September 12th, 2004*. Boston, 117–124.

Favareau, Donald 2007. The evolutionary history of biosemiotics. In: Barbieri, Marcello (ed.), *Introduction to Biosemiotics: The New Biological Synthesis*. Dordrecht: Springer, 1–67.

Forbus, Ken D. 1988. Qualitative physics: Past, present, and future. In: Shrobe, Howard (ed.), *Exploring Artificial Intelligence*. San Mateo: Morgan Kaufmann, 239–296.

---

[26] *Acknowledgements* go to Floyd Merrell, Claus Emmeche, Jesper Hoffmeyer, Donald Favareau, Günter Witzany, John Deely. An earlier version of this paper was presented at the 6th *Gathering in Biosemiotics* (Kull 2007b).

Hoffmeyer, Jesper 2008. From thing to relation: On Bateson's bioanthropology. In: Hoffmeyer, Jesper (ed.), *A Legacy for Living Systems: Gregory Bateson as Precursor to Biosemiotics*. Berlin: Springer, 27–44.

Ivanov, Vyacheslav Vsevolodovich 1978 [1962]. The science of semiotics. *New Literary History* 9(2): 199–204.

Kauffman, Louis H. 2005. Virtual logic — the one and the many. *Cybernetics and Human Knowing* 12(1/2): 159–167.

Kaushik, Meena; Sen, Aparna 1990. Semiotics and qualitative research. *Journal of the Market Research Society* 32(2): 227–242.

Kull, Kalevi 1992. Evolution and semiotics. In: Sebeok, Thomas A.; Umiker-Sebeok, Jean (eds.), *Biosemiotics: Semiotic Web 1991*. Berlin: Mouton de Gruyter, 221–233.

Kull, Kalevi 2002. A sign is not alive — a text is. *Sign Systems Studies* 30(1): 327–336.

Kull, Kalevi 2007a. Biosemiotics and biophysics — the fundamental approaches to the study of life. In: Barbieri, Marcello (ed.), *Introduction to Biosemiotics: The New Biological Synthesis*. Berlin: Springer, 167–177.

Kull, Kalevi 2007b. Life is many: On the methods of biosemiotics. In: Witzany, Günther (ed.), *Biosemiotics in Transdisciplinary Contexts: Proceedings of the Gathering in Biosemiotics 6, Salzburg 2006*. Salzburg: Umweb, 193–202.

Kull, Kalevi 2009. The importance of semiotics to University: Semiosis makes the world locally plural. In: Deely, John; Sbrocchi, Leonard G. (eds.), *Semiotics 2008: Specialization, Semiosis, Semiotics*. Ottawa: Legas, 494–514.

Kull, Kalevi; Torop, Peeter 2003. Biotranslation: Translation between umwelten. In: Petrilli, Susan (ed.), *Translation*. Amsterdam: Rodopi, 315–328.

Lotman, Juri 1990. *Universe of the Mind: A Semiotic Theory of Culture*. London: I. B. Tauris.

Maasik Sonia; Solomon, Jack 1994. The semiotic method. In: Maasik Sonia; Solomon, Jack (eds.), *Signs of Life in the USA: Readings on Popular Culture for Writers*. Boston: Bedford/St. Martin's, 4–9.

Manning, Peter K. 1987. *Semiotics and Fieldwork*. (Qualitative Research Methods 7). Newbury Park: SAGE Publications.

Markoš, Anton 2002. *Readers of the Book of Life: Contextualizing Developmental Evolutionary Biology*. Oxford: Oxford University Press.

Maturana, Humberto 1980 [1970]. Biology of cognition. In: Maturana, Humberto; Varela, Francisco J., *Autopoiesis and Cognition: The Realization of the Living*. Dordrecht: D. Reidel Publishing Co., 5–58.

Merrell, Floyd 2007. Toward a concept of pluralistic, inter-relational semiosis. *Sign Systems Studies* 35(1/2): 9–70.

Pattee, Howard H. 1972. Laws and constraints, symbols and languages. In: Waddington, Conrad Hal (ed.), *Towards a Theoretical Biology*, vol. 4, *Essays*. Edinburgh: Edinburgh University Press, 248–258.

Petitot, Jean; Smith, Barry 1990. New foundations for qualitative physics. In: Tiles, James E.; McKee, Gerard T.; Dean, C. Garfield (eds.), *Evolving Knowledge*

in Natural Science and Artificial Intelligence. London: Pitman Publishing, 231–249.
Petrilli, Susan 2009. *Signifying and Understanding: Reading the Works of Victoria Welby and the Signific Movement*. Berlin: De Gruyter Mouton.
Polanyi, Michael 1968. Life's irreducible structure. *Science* 160: 1308–1312.
Portmann, Adolf 1990. *Essays in Philosophical Zoology by Adolf Portmann: The Living Form and the Seeing Eye*. (Carter, Richard B., trans.) (Problems in Contemporary Philosophy 20.) Lewiston: Edwin Mellen Press.
Rosen, Robert 1991. *Life Itself: A Comprehensive Inquiry into the Nature, Origin, and Fabrication of Life*. New York: Columbia University Press.
Rosen, Robert 1999. *Essays on Life Itself*. New York: Columbia University Press.
Rosenthal, Sandra B. 1994. *Charles Peirce's Pragmatic Pluralism*. Albany: State University of New York Press.
Rothschild, Friedrich Salomon 1962. Laws of symbolic mediation in the dynamics of self and personality. *Annals of New York Academy of Sciences* 96: 774–784.
Rothschild, Friedrich Salomon 1968. Concepts and methods of biosemiotic. *Scripta Hierosolymitana* 20: 163–194.
Shank, Gary 1995. Semiotics and qualitative research in education: The third crossroad. *The Qualitative Report* 2(3).
Short, Thomas L. 1986. Life among the legisigns. In: Deely, John; Williams, Brooke; Kruse, Felicia E. (eds.), *Frontiers in Semiotics*. Bloomington: Indiana University Press, 105–119.
Stjernfelt, Frederik 1992. Categorical perception as a general prerequisite to the formation of signs? On the biological range of a deep semiotic problem in Hjelmslev's as well as Peirce's semiotics. In: Sebeok, Thomas A.; Umiker-Sebeok, Jean (eds.), *Biosemiotics: The Semiotic Web 1991*. Berlin: Mouton de Gruyter, 427–454.
Uexküll, Jakob von 1936. *Niegeschaute Welten: Die Umwelten meiner Freunde*. Berlin: S. Fischer.
Uexküll, Thure von 1992. Introduction: The sign theory of Jakob von Uexküll. *Semiotica* 89(4): 279–315.
Vehkavaara, Tommi 2002. Why and how to naturalize semiotic concepts for biosemiotics. *Sign Systems Studies* 30(1): 293–313.

Part II
APPLICATIONS

Chapter 7
The Need for Impression in the Semiotics of Animal Freedom: A Zoologist's Attempt to Perceive the Semiotic Aim of H. Hediger

Aleksei Turovski

Summary. The works, views and ideas of Heini Hediger (1908–1992) had and still have an enormous impact on understanding of animal behaviour. His views on territorial, social, etc. aspects of animal behaviour are based on semiotic concepts derived from Uexküll's umwelt theory and combined with ideas from modern ethology. Hediger's special attention was devoted to the area of animal–human communications; he treated these problematic phenomena as a system of semiosic processes, in a mainly holistic way. Hediger's approach inspires the author to propose the notion, "the need for impression" to be used in zoosemiotic analyses.

> *My childhood dream, my lifelong wish, would have been fulfilled if it had really been possible to converse with animals.*
> H. Hediger (1985: 177)

> *There's a disadvantage in a stick pointing straight (because) the other end of the stick always points the opposite way. It depends whether you get hold of the stick by the right end.*
> G. K. Chesterton (1984: 221)

There have never been any difficulties in pointing out the signs of captivity: the bars and gratings, the fetters and shackles, the nets and traps–signs of obstacles, hindrance and limitations of any kind will certainly do. But it is not at all easy to find a universally acceptable and generally comprehensible symbol even of human freedom, so what could serve us as a sign of animal freedom? The image of an animal moving freely is a sign of escape rather than that of a free life *par excellence*. The general notion of a free animal has always been something of vague perception of a completely undetained creature, imperceptible though certainly dwelling somewhere in the wild.

It is hard to believe in our time, but it seems that Heini Hediger (1908–1992) really was the first zoologist who realized that there is no such thing as an animal that is free in an anthropomorphic sense: "the free animal does not live in freedom: neither in space nor as regards its behaviour towards other animals" (Hediger 1964: 4).[1] Animals in the wild are "bound by space and time, by sex and social status" (Hediger 1985: 158). If we consider now that Hediger also elaborated on the distinction between *nest* and *home*, the former being "a repository for eggs and raising the young", and the latter "a place of refuge, which is the function of the home" (Hediger 1985: 178), it becomes quite clear, for a zoologist at least, how Hediger approached the phenomena of animal life. He did it from within, treating the living world as the umwelt in Jakob von Uexküll's sense.

To Uexküll Hediger dedicated his study on tameness (Hediger 1935).[2] He was a friend and admirer of Jakob von Uexküll and Thomas A. Sebeok, but was he a semiotician? As far as I know, he was a zoobiologist and field zoologist in the first place, deeply involved in zoopsychological and behavioural studies, friend and adherent of K. Lorenz and N. Tinbergen, though not an ethologist proper (Hediger 1985: 178–179). Having enormous experience in the ecology of animal behaviour both in the field and in captivity, Hediger doubted the somewhat rigid interpretations of the instinct,

---

[1] On Hediger's biography and work, see Sebeok (2001b); Honegger (1993); Hediger (1985).

[2] About this fact see Hediger (1985: 149).

the ritualistic behaviour in the first place, adopted by ethology as a science of species-specific behaviour of animals. Indeed, the vast diversity of deviations from the main schemes of instincts within one and the same species, the flexibility and continuity of the adaptive behaviour of animals, give us reason to believe that the individual organism itself is the most active interpreter of the innate mechanisms of behaviour, and that the main matrix on which these interpretations evolve during the ontogenesis is the semiosphere. Hediger distrusted "the accepted evolution theory", not accepting the claim that the "two major constructors of speciation are mutation and selection" (Hediger 1985: 179). And Hediger certainly was not a behaviourist in classical sense. It is my conviction that, in all his studies, especially "The Clever Hans phenomenon from an animal psychologist's point of view" (Hediger 1981), Hediger operated with conceptual notion-instruments *signum–structure* in the cases where behaviourism would apply the famous *stimulus–reaction* scheme instead. In all his studies and encounters with animals in the field and the zoo Hediger's attention was driven to the biological meaning of the signal — an impulse of information passing between animal and other components of the umwelt, including all human factors as well. This reminds me of what my father Markus Turovski, a philosopher, once said to me about his attitude to living things: "If I were told that an octopus can talk in, say, English or Russian, I would consider it simply as a fact of its personal biography. What it is talking about is the only thing that matters to me."

Thomas A. Sebeok describes Hediger as a "visionary innovator who reached from the inside outwards. He felt entirely comfortable within Jakob [von Uexküll]'s umwelt paradigm, but, implicitly, with (zoo)semiotics too, which he came increasingly and quite explicitly to embrace" (Sebeok 2001a: 67). Sebeok highly values Hediger's works on territorial and social behaviour, especially his concepts of "individual distance" and "home range" together with distinguishable "territorial idiolects and dialects" as characteristics for the communicative patterns of the species sharing a home range, and also Hediger's views on hierarchy and dominance in animal social status, on parental care and other forms of communicative stimulation/inhibition activities. In these concepts as well as in all his logically extremely coherent works, Hediger is estimated by Sebeok as a true zoosemiotician whose works offer materials and ideas of great importance for semiotics of all scientific trends — anthropological, perhaps, in the first place (Sebeok 1972: 172–173; 1989: 5, 55; 1990: 107, 124).

Hediger, though he probably did not call himself a semiotician, obviously worked like one. And following the aim of his works, we could describe the free animal as a representative of a particular species, active in its specific semiosphere as part of the umwelt. The contacts (and conflicts) between an animal and human culture, and furthermore between its species and civilization could be then understood as interactions between animal and human semiospheres, so the main aspect of these interactivities is the dynamics of attention on both sides.

In "Communication between man and animal" (Hediger 1974), "Man as a social partner of animals and vice-versa" (Hediger 1965), "Wild Animals in Captivity" and many other studies, Hediger emphasizes the absolute necessity of understanding the actual animal–man encounter situations by their signs as examined from animals' point of view in the first place, in order to find optimal means of control. "People never answer what you say. [...] They answer what they think you mean" (Chesterton 1984: 76). Actually animals act in the same way and it is, essentially, up to the inquisitive interrogator — the man, to arrange "the questionnaire" by meanings, that is, to formulate it semiotically in elaborated sets of signs, fitted in to the space–time structure of the semiosphere of particular animal species. In the course of a lifetime, that is in embryonic, postnatal, juvenile, subadult, adult/imagial-sexually mature and postmature periods, an animal passes through a succession of very different behavioural stages of orientation from other animals (and man) as objects of its attention. One of the important attitudes in this process

The Need for Impression in the Semiotics of Animal Freedom 137

besides the motivational attitudes to resources, foes, sexual partners, social ranks, broods etc., is the fulfilment of the need to be impressed by changing signals — impulses of information from the environment, otherwise indifferent in the aspects of major biological needs/functions. The matrix structure of the semiosphere of the animal obviously transforms these signals into signs in accordance with the prevailing motivations; so the forms of the umwelt become semiotically involved in the unique personal experience of the animal in dependence on its ontogenetic age-period. Such a "need for impression" is presumably coupled with the need to impress and thus to provoke feedback signals which also contribute to the process of semiosis. Man has been making use of the need for impression, calling it "natural curiosity" of animal, in taming and domestication.

Apparently there is no such thing as a "free population", to say nothing of "a free species". But a free animal could, perhaps, be usefully imagined as a healthy member of a healthy population of a certain species dwelling in such part of the land- or seascape which is safe from foes, but promising in resources as far as this animal can recognize by its semiotic means within the range and limitations of its sensorics. Or, to make a long story short, free is an animal for which the fulfilment of the need for impression is granted by the functional structure of the semiosphere.

For a free animal, the best way to avoid encounters with all and any possible enemies is assumably to make itself imperceptible, to dissolve semiotically into its habitation, become undetectable. But this can only be effective until the event of disclosure happens. From this moment on the next necessary and diligent course of action is either to flee or, if escape is impossible or of dubious effect, to kill or scare the enemy off, or to gain time by stunning it, making it hesitate even for a moment.

Presumably, the best way to do this is to produce such a complete set of the signs of danger as to force every particular enemy to choose by itself the most horrifying pattern in accordance with its specific properties and personal experience. That can be done to any animal species because all forms of animal life are united semiotically by the **need for impression**.

The basic importance of the need to achieve and to produce the impression could be illustrated by the quite general and indeed universal pattern in choosing a leader (and a sexual partner in females) in social behaviour of animals. The decisive factor in those processes is *charisma* as the ability of a subject to perform a display of impressive signs of irresistible dominativeness and to sustain such an impression among the members of the group for as long as possible. The leader must never and on no account display in any situation any signs of hesitation.

Quite convincing evidence of the importance of the need for impression in humans could be easily found by an approximate estimation of the number of horror film addicts. In the well-known theory of mimicry, the most active, essential role belongs to the "dupe". The "dupe" is the actual decider on the course of events and final results in all situations where camouflage, mimicry and other such phenomena are involved. So, it is the "dupe" to whom the set of repelling patterns of form and behaviour is offered to make its own choice what particular sign to be scared of. And it is ecologically essential that the life of the dupe must be preserved (e.g., Mertensian mimicry clearly shows in the case of deadly coral snakes that they are mimics, but not models), for a dead dupe can neither learn, nor pass the knowledge. Semiotically man holds the unique position in the umwelt, viewed in the aspect of mimicry in "hide, seek, scare and catch" games, because man could rationally take on any or all of these roles at the same time, avoiding being duped. That gives humans the most advantageous position as hunters, tamers, domesticators, and most perilous exterminators as well. Unless I am very much mistaken, such was the main concern of Heini Hediger as a zoologist, zoo director and a man to whom communications with animals and zoological studies were at least equally important. This dualistic attitude could probably be marked as romantic, but it certainly is semiotic.

If I may be allowed a lyrical reminiscence, Hediger's inclination to approach life problematics "from within" reminds me of Geheimer Archivarius Lindhorst, actually a Salamander (Fiery Spirit of Nature), the most powerful character in Hoffmann's "Der goldne Topf" (Hoffmann 1814), whose principle was to act always "from within out" in order to restore the harmony between man and nature.

All my personal experience obtained in 38 years with animals in Tallinn Zoo indicates clearly that the main and very first concern of an animal that finds itself outside its enclosure is to ensure the safest and straightest way back home. This home, as is pointed out above, is in Hediger's sense a place for refuge. But to recognize the **home from outside** could be extremely difficult for an animal (and also for a human as it happens in various situations), accustomed to experiencing the world strictly **from within**. So the animal is always most grateful (or at least we can say relieved) for any assistance from humans (or other animals, e.g., dogs), which are considered by the animal as its conspecifics, if the animal is tamed or properly acquainted with man. Generally speaking, humans could find themselves in a very similar but reversed situation in the near future. Until recent years the concept of civilization for the majority of at least Western Europeans has obviously been based on the conviction that the landscape is something extremely stable and unchanging "by itself". Now, with global warming and other such troubles it seems to be possible that people would find themselves on the "outside" of their home and with anxious expectations that just animals will "show the way back inside". I think that such would have been H. Hediger's feelings now.

As for Hediger's doubts about the synthetic theory of evolution, those I think were based mainly on such observable events as the preference given by females to childlike or otherwise deviated males instead of very masculine ones, the general infantilization and sexual acceleration in zoo animals (and others), on the striking importance of imprinting combined with games and teaching in raising the young in predatory mammals and birds (but also some fish and even insects, e.g., *Passilidae*, *Coleoptera*), and on the vast variability in communication patterns in animal behaviour. But, as all the scientific legacy of Hediger, this topic needs further study. Reproductive isolation may occur within a population due to behavioural particularities of single organisms. It is well known that in some cases females prefer males clearly deviated from the median pattern of behaviour, e.g., wasps *Mormoniella vitripennis* (White, Grant 1977), *Drosophila* sp. in the phenomenon of asymmetric evolution on Hawaii islands (Lambert 1984), though, as a rule, female mate choice favours symmetrical males in such different species as barn swallows (*Hirundo rustica*), earwigs (*Forficula auricularia*), humans (*Homo sapiens*) and many others (Polak 1997). For my part, the study of the possibilities to apply the Baldwin principle in the analyses of zoosemiotic views of Hediger seems to be very promising. In this field it would be especially interesting to try the application of K. Kull's ideas on the Osborn–Baldwin effect (Kull 2000). According to his interpretation of this effect, "the activity of an organism as a subject may play a role as an evolutionary factor" (Kull 2000: 53).

The notion "activity of an organism as a subject" in the case of an animal implies a concept of behaviour as an adaptive holistic system of interconnected *choices of the course of action* (resembling, perhaps, the "creodes" in the Waddington sense) *in the semiosphere* (as in the field of signs). The adaptability of such a system assumes the ability of episodic and situational memory combined with a set of emotions viewed as innate mechanisms of mobilization and direction of behavioural energy. In this discourse the behavioural activity of an animal is closely related to subjective experience, individual as well as social. Thus the subjective activity based on experience constitutes a trend of functional involvements of semiotic traits in evolutionary processes.

One must not expect to have a powerful and beautiful waterfall without the river. Such expectations can never arise where the umwelt theory is semiotically correctly applied as it was always done by Dr. Heini Hediger in all his studies.

References

Chesterton, Gilbert Keith 1984. *The Penguin Complete Father Brown*. London: Penguin Books.
Hediger, Heini 1935. Zähmung und Dressur wilder Tiere. *Ciba Zeitschrift (Basel)* 3(27).
Hediger, Heini 1964. *Wild Animals in Captivity: An Outline of the Biology of Zoological Gardens*. New York: Dover Publications.
Hediger, Heini 1965. Man as a social partner of animals and vice-versa. *Symposia of the Zoological Society of London* 14: 291–300.
Hediger, Heini 1974. Communication, between man and animal. *Image Roche (Basel)* 62: 27–40.
Hediger, Heini 1981. The Clever Hans phenomenon from an animal psychologist's point of view. *Annals of the New York Academy of Sciences* 364: 1–17.
Hediger, Heini 1985. A lifelong attempt to understand animals. In: Dewsbury, Donald A. (ed.), *Leaders in the Study of Animal Behavior: Autobiographical Perspectives*. Lewisburg: Bucknell University Press, 144–181.
Hoffmann, Ernst Theodor Amadeus 1814. *Fantasiestücke in Callots Manier*, Bd. 3. Berlin: Kunz Verl.
Honegger, René E. 1993. Heini Hediger (1908–1992). *Copeia* 2: 584–585.
Kull, Kalevi 2000. Organisms can be proud to have been their own designers. *Cybernetics and Human Knowing* 7(1): 45–55.
Lambert, David M. 1984. Specific-mate recognition systems, phylogenies and asymmetrical evolution. *Journal of Theoretical Biology* 109(1): 147–156.
Polak, Michal 1997. Parasites, fluctuating asymmetry, and sexual selection. In: Beckage, Nancy E. (ed.), *Parasites and Pathogens: Effects on Host Hormones and Behavior*. New York: Chapman & Hall, 246–276.
Sebeok, Thomas A. 1972. *Perspectives in Zoosemiotics*. The Hague: Mouton.
Sebeok, Thomas A. 1989. *The Sign and its Masters*. London: University Press of America.
Sebeok, Thomas A. 1990. *Essays in Zoosemiotics*. Toronto: University of Toronto.
Sebeok, Thomas A. 2001a. Biosemiotics: Its roots, proliferation, and prospects. *Semiotica* 134(1/4): 61–78.
Sebeok, Thomas A. 2001b. *The Swiss Pioneer in Nonverbal Communication Studies Heini Hediger (1908–1992)*. New York: Legas.
White, Harry C.; Grant, Bruce 1977. Olfactory cues as a factor in frequency-dependent mate selection in *Mormoniella vitripennis*. *Evolution* 31(4): 829–835.

Chapter 8
The Multitrophic Plant–Herbivore–Parasitoid–Pathogen System: A Biosemiotic Perspective

Luis Emilio Bruni

Summary. In the past three decades there has been an increasing number of studies concerned with the effects that alterations in biodiversity may have on ecosystem functioning. In these studies a great emphasis has been on ecological processes such as productivity, energy flow and nutrient cycling. The models for multitrophic interactions and above and below ground interactions have mainly furnished a picture based on material exchanges, i.e., trophic webs, between the participating taxa. On the other hand, in most disciplines of biology there is an incipient trend that considers biology as a science of "sensing", that is, biologists in different sub-disciplines are assigning increasing importance to the informational processes in living systems and are paying more attention to the "context" (e.g., from quorum sensing to info-chemicals to signal transduction in general). There is a new and exciting epistemological path opened in ecology which is seriously considering the evolution of signals as one of the most important processes in ecosystems functioning. There is in the literature a call for integration of molecular and ecological perspectives. But instead we find a tendency to reduce the latter to the former, that is, to decompose (reduce) the ecological complexity into its molecular "components". The aim of this chapter is to present the biosemiotic perspective as a useful conceptual framework to reorganize and interpret the general model for ecosystem functioning (in relation to biodiversity changes) present in many different empirical studies of what we could call the "multitrophic plant–herbivore–parasitoid–pathogen system".

Introduction

Several different concepts have been used during the 20th century to represent the totality of living nature. Consider for example two of the most accepted, namely the biosphere and the biodiversity concepts. These two notions constitute different approaches to biological complexity, although they have in common the fact that they focus our attention upon a network that may include everything from genes to ecosystems. While biodiversity has been understood in terms of its "components", for the most part ignoring the relations between them, the biospheric approach (biogeochemical aspects of ecosystems) has surrendered to the strategy of explaining life as "nothing-but-interacting-molecules", resulting in an explanation of life as trophic chains and mass-energy exchanges at ecosystem level.

Today the interest is shifting increasingly towards a combination of both points of views, i.e., the relation between ecosystem functioning and biodiversity changes or, conversely, the functional role of biodiversity in ecosystems processes. According to Haber (1999: 179) it is preferable to study "small" ecological systems or subsystems, like plant–insect complexes, or communities in the smallest aquatic or terrestrial habitats, and he asserts that what assures that biodiversity is adequately taken into account in this type of research is a combination of the concept of functional groups/ecological guilds and the study of food chains and food webs. In fact, a variety of empirical studies have yielded a general model that could be called the "multitrophic plant–herbivore–parasitoid–pathogen system". These studies, which increasingly consider above and below ground multitrophic interactions, until recently had furnished a picture based on material exchanges, i.e., trophic webs, between the participating taxa.

Haber (1999: 179) further suggests that future investigations must focus on interactions between species, in two different ways: (1) on the interactions between the species themselves (which would be the task of biocoenotic research); and (2) on the interactions or connections of the species with their abiotic environment (which is "classical" ecology). In our view it would be a great mistake to keep these two aspects separate. Both kinds of interactions determine the *context* in which organisms are immersed and synergically constitute what has been defined as the "semiotic niche" of the species (Hoffmeyer 1997).

According to Kratochwil (1999), there is no doubt that many organism species are constantly linked by certain interactions, and that these interactions may be obligatory. But he then adds that such an interaction structure only has system character when it can be differentiated from other systems and when an independent matter flow is ascertainable. In other words, it is normally claimed that interactions amount to matter (nutrient) and energy flow. In fact, Haber (1999: 176) points out that the great biological and ecological research programs of the last decades focused mainly on the discovery or confirmation of universal natural laws, predominantly on absorption, transformation, and processing of energy and matter. But in the last decade we can also observe a trend that considers biology as a science of "sensing" in which biologists from different sub-disciplines are assigning increasing importance to the "informational processes" in living systems and paying more attention to the "context" (Bruni 2003; 2007; 2008).

Kratochwil (1999: 14) quotes Günther's definition of niche, according to which the niche is a dynamic relation system of a species with its environment. It is composed of an autophytic/autozooic and an environmental dimension. The autophytic/autozooic dimension comprises the phylogenetically acquired morphological and physiological characteristics of the species while the environmental dimension is the sum of all effective ecological factors. However, Günther's definition (dating from 1950) states that the autozooic dimension (in the case of animals) also comprises ethological characteristics. This consideration makes the issue of interactions much more complex than just matter and energy exchanges and it does in fact introduce what I will be calling the "semiotic dimension". To this we can add that today it is very common among botanists to assume an "ethological dimension" for plants as, for example, when they talk about behavioral traits, not to mention communication between plants.

Haber (1999: 179) claims that biological diversity is often understood exclusively — or mainly — as diversity of tangible structural entities. This state of affairs gives rise to a huge knowledge gap about the diversity of "ecological niches", i.e., of relations between organism and environment. This means that there is very little consideration of the surprising variety of communication systems within species and populations, as well as of the diversity of behavior expressions and learning processes. Clearly what is lacking is the semiotic dimension.

There is in the literature an increasing interest in "complex interactions", "multitrophic links", "connectivity", responses to "multiple enemies", "cross-talk" between multiple pathways, "non-trophic interactions" relations between biotic and abiotic factors, between above and below ground, between ecological and historical developments. What does this mean epistemologically? It looks as if after making inventories of components and dissecting individual pathways, the reductionist strategy is in need of a complementary perspective to "connect" all the reduced parameters which in the "field" influence each other, creating cocktails of non-linear causal links.

The reductionist strategy has concentrated on the accountability of material stocks (whether in terms of matter or energy). But we have seen in the last decade a renewed interest in the "flow of information" within these complex processes. We find terms such as "semiochemicals", "chemical information", "signals", "sensing", "recognition" and "perception" (see for example: Vos *et al.* 2006; Ohgushi 2008).

Having research dominated by the reductionist approach, it is no surprise that with the rapid development of molecular techniques there is in fact a sort of interaction of all branches of biology with molecular biology, including ecology. In this regard there is also in the literature a call for integration of molecular and ecological perspectives (Baldwin *et al.* 2001; Paul *et al.* 2000). How are these approaches being integrated? Are they really being integrated? Or is it just a reduction of the latter to the former? With the importing of the concern for "information" from molecular biology to ecology, we may also be importing the ambiguity that has characterized the "informational talk" in 50 years of molecular biology. Yet it may very well be that the element that may bring about the longed-for integration across hierarchical levels and sub-disciplines is precisely the "information" notion, or better yet, the semiotic dimension.

Biological information

From a biosemiotic perspective there is a lot of interest in clarifying the notion of "biological information" at the different hierarchical levels of biology: from the molecular-genetic level to the epigenetic (whole cell) level up to more systemic levels that include various types of communication systems such as nervous, immunologic, endocrine and ethological systems, up to ecosystems. Here the emphasis is not merely on the "transfer of information" per se, as if it was a material thing that can be physically moved from one place to another (whether in genetic or in ecological systems), but on the emergence of communication webs and interpretation contexts and systems.

Besides a need to integrate molecular and ecological approaches when investigating the role of above–belowground multitrophic interactions of plants, herbivores, pathogens, and their antagonists, we can also find a call for new theoretical and empirical investigations in the literature (Putten *et al.* 2001: 553). If one considers the growing field of research around the notions of info- or semiochemicals, which imply a sort of "transfer of information", the biosemiotic considerations on "biological information" could be of help in the theoretical endeavor.

According to Van der Putten *et al.* (2001: 553) "[...] new studies should consider the roles of plant nutrition, nutritive compounds, signal transduction pathways and spatio-temporal processes". It looks as if signal transduction networks constitute one of the first conceptual links between molecular and ecological approaches in general, and specifically regarding ecological studies of above–belowground multitrophic interactions (Bruni 2003; 2007).

While on the one hand the integration of molecular and ecological approaches is more than necessary, there is yet the risk of importing some of the ambiguities of the "informational talk" from molecular biology to ecology. So it would be recommendable to accompany the integration with a concomitant clarification of the information concept and its ontological and epistemological implications. The central assumption of this chapter is that instead of originating exclusively at the molecular-genetic level, "biological information", as the vehicle for communication, must present common features and causal relations at all different levels and systems.

Biology becomes a science of "sensing"

The biosemiotic approach has repeatedly emphasized that in the last three decades there has been in biology a shift from a focus on information (as a material agent of causality) towards a focus on signification processes, i.e., information as context-dependent. From different directions we can corroborate that biology is becoming a "science of sensing" (e.g., from quorum sensing to info-chemicals to signal transduction in general) and that there is an increasing importance being ascribed to the "context" in experimental biology.

Quorum sensing research may be a paradigmatic example of this trend. "Quorum sensing" refers to one of the many transcription regulation systems in prokaryotes, one which is coupled to intercellular communication mediated by signal molecules that are thought to constitute inter-bacterial communication codes. The dynamics involved in the evolution of these phenomena represent an interesting instance of emergence of informational contexts along the biological hierarchy from molecules to ecologies. The complex specificity of the host-symbiont relation is determined by an aggregate set of specificity determinants which in turn are conformed by lower level specificities (Bruni 2002).

It must be emphasized that the inter-bacterial code active in quorum sensing may be an integral part of the semiotic network operative in the "above–belowground multitrophic plant–herbivore–parasitoid–pathogen system" since almost all cases of quorum sensing are related to symbiosis. Nearly always the colony that coordinates the simultaneous expression of a given phenotype is a symbiont of a higher organism and very often the cell-density-dependent phenotype is related to the colonization and/or the interaction with the host.

There are several lessons that quorum sensing research can teach us when considering networks of info-molecules in multitrophic systems. As in the 1990s the list of bacteria that possessed quorum sensing systems expanded, and so did the list of phenotypes regulated in this manner, and the family of homoserine lactones that serve as signal molecules (Salmond *et al.* 1995; Swift *et al.* 1999), researchers wondered how did these signals evade detection for so long admitting that such exchange of external signaling molecules between single celled organisms was unexpected and therefore nobody was looking for them. For decades, microbiologists had been isolating cells out of the culture medium in which they had grown and throwing that medium away together with the signals. That is why some bacteria would lose their pathogenicity in the experimental settings. It was the *context* that was being thrown away!

A narrow "struggle for life" view may lead us to think that every time we encounter a so-called antibiotic in nature we have before us a case of biochemical warfare. Perhaps this is not something we should take for granted. As in symbiosis in general, there are plenty of examples of mutualistic interactions via quorum sensing, not only in the symbiosis bacteria-higher organism, but also in bacterial interspecies communication, or cross-talk. There is also evidence that some bacteria may become virulent in response to cell-signals from quite unrelated bacteria in the environment, and different species have been reported to team up and communicate in order to coordinate their pathogenic response (Eberl *et al.* 1999: 1708–1710). This simply means that any assessment of an organism's virulence must take into account the context and the likelihood of signaling molecules being present, i.e., assessment of the semiotic niche. There are many peptide pheromones and other bacterial signal systems that cross-talk commonly being reported. Certain cross-talking signals have also been identified in biological systems as different as bacteria and mammals (e.g., cyclic dipeptides found in marine bacteria have been found in mammalian systems such as neurotransmitters) (Rice *et al.* 1999: 28).

From semiochemicals to ecosystems

In a "multitrophic" system, which is simply what reality offers (since the bitrophic and even the tritrophic systems are artificial realities), there are multiple interrelations that create many possibilities for combinations of mutualisms, amensalisms, parasitisms, predation, comensalisms and antagonisms. For example, in a terrestrial multitrophic system bacteria can antagonize with a fungal "pathogen" (isn't the bacteria pathogenic to the fungus?) which in turn can antagonize with the host-plant, the bacteria being a mutualistic symbiont of the plant (Seddon *et al.* 1997; Keel, Défago 1997). Soil microbiota determine many types of interactions in the rhizosphere including the so-called plant growth promoting rhizobacteria, symbiotic nitrogen-fixing bacteria and mycorrhizal fungi that constitute a mycelium "bridge" connecting the plant's roots with the surrounding soil microhabitats and to other plants of the same or different species (Barea *et al.* 1997). We encounter fungal — fungal interactions (those considered pathogenic and those considered beneficial interacting between them) (Whipps 1997); interactions between mycorrhizal fungi and foliar fungal pathogens in the phyllosphere of host plants (West 1997); host-plant mediated interactions between arbuscular mycorrhizas and plant parasitic nematodes (Roncadori 1997), between arbuscular mycorrhizas and subterranean and foliar-feeding insects (Gange, Bower 1997); between insect herbivores and pathogenic fungi on the phyllosphere (Hatcher, Ayres 1997); between micro-herbivores (such as bacteria, fungi and viruses) and macro-herbivores (invertebrate and vertebrate) (Faeth, Wilson 1997); to name just a few cases recently under investigation.

These systemic interactions also include temporally and spatially separated species interactions (Faeth, Wilson 1997: 202). The experimental approaches consider both top-down effects (control exerted by predators on lower trophic levels) and bottom-up effects (control exerted by resources available to each trophic level). The consideration of several processes acting simultaneously (predation, disease, competition for resources, competition for enemy-free space, limitations imposed by abiotic conditions, etc.) has been recommended (Karban 1997: 199).

The material exchanges of these complex interactions (nutritional quality of tissues, nutrients and metabolites concentrations, etc.) have been the subject of many studies and are well documented. On the other hand the "non-trophic" relations, which may not be as easily discernible as trophic interactions, are just recently being characterized. There is no doubt now that non-trophic interactions play an equally important role in ecosystem functioning as trophic interactions do. The nature of these interactions hints to communication (semiotic) processes. As pointed out by Putten *et al.* (2001: 548), "The information used by aboveground invertebrate herbivores and carnivores overlays the food web and includes cues to factors that mediate indirect interactions between species." They further add that "Food-web models could be useful for evaluating the consequences of above-belowground links of multitrophic interactions.

Although these models are based mainly on flows of energy and nutrients, they could be extended to account for spatial and temporal heterogeneity. Some organisms, such as parasitoids and pathogenic fungi, might strongly affect the stability of food chains, whereas they make very little direct contribution to flows of energy or nutrients" (Putten *et al.* 2001: 553).

Baldwin *et al.* (2001) suggest merging molecular and ecological approaches. However, they do not advance any theoretical suggestion that could integrate the different hierarchical levels. They review the molecular details of signaling processes and point at some of their ecological consequences. But instead of merging both hierarchical levels, molecular and ecological (and one should also consider the epigenetic and developmental continuum in between), this may tend to reduce the latter to the former, that is, decompose (reduce) the ecological complexity into its molecular "components", with the understated goal of mapping an ecosystem in terms of molecular kinetics. While this approach is indispensable, given that semiotic processes operate through such a complex substrate of interacting molecules, it can not be expected to exhaustively account for the emergence of novel semiotic hierarchical networks that give rise to systems of semiotic control over the flux of metabolites.

So if we have on the one hand the complexity implied by the consideration of multitrophic interactions and the reciprocal influences of what happens — not only above and below ground — but also laterally in the whole periphery of the niche or even in the whole ecosystem, we have on the other hand that this complexity increases further when we consider the plethora of non-trophic interactions. Here the elemental "mechanism" or "action" is the ability of organisms to sense, integrate and exchange molecular signals with a myriad of beneficial and harmful organism, i.e., sophisticated molecular mechanisms to respond to and to communicate with, for example, pathogens and symbionts (Staskawicz, Parniske 2001: 279).

Commenting upon a study that hints at the involvement of herbivores' gut bacteria in the production of the elicitors that plants use to recognize the herbivores and call for help from the third trophic level (the herbivores' predator), implying therefore a fourth interaction level, Baldwin *et al.* (2001: 355) claim that this complexity is likely to increase when the influence of mycorrhizae and endophytes are considered. To approach this complexity it will be necessary to consider the intrinsic relation between endo- and exosemiotic codes (Bruni 2007). Baldwin and his colleagues further claim that an intimate understanding of the ecology of the plant system (in their case *Nicotiana attenuata*) involved in defense against insects is necessary to decipher the transcriptional Rosetta Stone. "Plant–insect interactions are played out in an ecological arena that is larger than the plant itself and incorporates many community-level components, as indirect defenses so clearly illustrate. These higher order interactions can reverse the fitness outcome of a trait" (Baldwin *et al.* 2001: 353).

Dicke and Bruin (2001: 988) argue that in the study of chemical information transfer between plants, much can be learned from research on chemical information in interactions between animals. In this regard they point to the growing body of evidence that animals exploit many sources of information (on resources, on competitors, on natural enemies) to adjust their behavioral decisions. In this context it is very tempting to reduce information to "chemical information" as it happened before with "genetic information", where information became DNA, i.e., matter. Maybe our tendency to give priority to "chemical information" over other kinds of cues and regularities is that it can give us the illusion of a material exchange and thus it simplifies the ambiguities of a not well explained information notion, if it does not ignore them altogether.

Dicke and Bruin (2001: 988) also remark that although interesting in itself, the medium of communication is of course not the main topic if one asks whether communication between damaged and undamaged plants occurs at all and how this affects the ecology of plant–attacker interactions. In fact, they notice the underground transfer of information may be facilitated by root networks and by mycorrhizal connections that may transport nutrients and potentially also elicitors of defense over considerable distances. As I interpret this remark, the main topic is the emergence of the network and the interpretation context. In other words, the "medium of communication" is only one of the three aspects that constitute something that we may call "information". Besides the "medium" (the sign), for information to be information there must also simultaneously exist that to which the sign refers (its "meaning") and the interpretation key, which completes the triadic relation, giving rise to the interpretation system to which the difference (created by the "medium") makes a difference.

Marcel Florkin in his treatise from 1974 "Concepts of molecular biosemiotics and molecular evolution", one of the early attempts to integrate a view of nature as a sign-producing system into an elaborated theoretical frame (Emmeche, Hoffmeyer 1991), recognizes the signified (that to which the sign refers, its "meaning") of biomolecules as being involved at levels of integration higher than the molecular one, for instance the level of self-assembly in supramolecular structures, and the physiological and ecological levels. He designates as "ecomones" the non-trophic molecules contributing to insure, in an ecosystem, a "flux of information" between organisms. I propose here to adopt Florkin's denomination of "ecomone" as a unified term that includes all instances of pheromones, infochemicals, semiochemicals, signals and non-trophic molecules present in ecosystems.

Usually in biology, when evolutionary considerations are invoked, what we are really interested in is what the relevance of a particular phenomenon, function, trait, or behavior is. Dicke and Bruin (2001: 988) assert that apart from mechanistic questions, evolutionary questions should be addressed asking why plants do (or do not) exploit their neighbor's information and whether their strategy is affected by e.g., environmental conditions or previous experience. This is equivalent to saying that the relevance and consequences of dyadic (mechanical) relations, i.e., their "significance", has to be seen within a triadic logic (Bruni 2003). As much as plant-to-plant communication is the last frontier being explored in interorganisms' communication, this is often investigated for interactions between conspecifics. Dicke and Bruin (2001: 988) see no good argument why plants would not be able to exploit chemical information from heterospecific damaged plants.

These mechanisms are based on a series of "recognition specificities" that apparently begin at the genomic levels of the interacting organisms (Jones 2001: 281). By analyzing the evolutionary aspects of these interactions it is hard to avoid noticing the coevolutionary nature of such communication "mechanisms". A clear example comes from the existence of multiple resistance specificities (determined by recognition specificities) that can occur as different haplotypes (set of alleles) in resistance (R) gene loci in a plant. A single gene in a given host can alternatively encode different "recognition specificities" that match different avirulence genes in a respective parasite (it is also hard to avoid noticing the parallelism with mammalians' immune systems). In other words, the plant's genome can recognize (through different binding specificities) sorts of "warning" signals encoded in the parasite's so-called avirulence genes, and thus initiate the resistance response.

There has been a lot of interest in the molecular and evolutionary mechanisms that create and sustain such diversity of recognition specificities (Jones 2001: 281). Two orthologous resistance genes encoding proteins with up to 90% amino-acid sequence identity can have distinct recognition capacities binding to two different corresponding avirulence proteins. This means that just a few amino-acids may be enough to confer the required specificity (Jones 2001: 282). In other words there are particular domains that make the major contribution to the unique recognition capacity of individual R genes. However, in some cases it may not be clear cut when a given protein should be considered as an avirulence or as a virulence factor, that is as a communication device (a sign for the potential victim) or as a weapon, and in some cases the same protein may play both roles (Nimchuk *et al.* 2001: 288). What determines the difference then must be the context, i.e., the presence of further simultaneous information that conveys a larger analogical message (Bruni 2007). This is what is implied when we repeatedly read about the influence of biotic and abiotic environmental conditions and of previous experience. This is also probably the main source of discrepancy between experimental and natural systems: the context.

Once more we find ourselves surrounded by words like communication, sensing, recognition and perception. The word "information" usually implies more or less similar albeit not well defined meanings. But in this new context, what is it that an organism can communicate? What is it that can be recognized, sensed or perceived? What can be sensed from the environment? And more importantly, who is the subject of these actions? For example, Paul et al. (2000: 221) point out that current theories on the evolution of induced defense are based on the concept that current herbivore or disease is correlated with the risk of future attack. In this sense, they claim, current or past attack can be seen as having the potential to provide information about the future environment. On this basis, induced resistances will be selected only if (i) current attack is a reliable predictor of future attack, and (ii) if attack reduces plant fitness. Here, "information" is the correlation of past or current attack with future attack. Are not we talking about learning and memory?

What is the information "conveyed" by current attack? But more importantly, information conveyed to whom? How and who knows that current attack "predicts" (it would be more precise to say "indicates", signs) the future attack? There is clearly an interpretation context involved here and current attack *per se* constitutes no information. It becomes information only in relation to an emergent interpretant (Bruni 2003; 2007). Contrary to mechanical actions and reactions, information requires a triadic logic. It involves not two but three elements in simultaneity constituting what in semiotics is referred to as the triadic sign-function. Information (current attack) is something that stands for something else (future attack) to some system with interpretation capacity, a subject (the lineage through the aggregate of individual plants) capable of sensing (recording, recognizing) the difference created by the event of the current attack, i.e., the reduced plant fitness. Current attack is information only if seen in this triadic relation.

In Paul *et al.* (2000: 224) there is reference to the relationships between plant responses to biotic and abiotic stress. The abiotic environment also provides "information" about the risk of future herbivores or disease. Responses to biotic and abiotic stress are linked because optimizing induced responses to minimize the physiological effects of attack is highly dependent on the abiotic environment.

We can clearly see here the context-dependent nature of any kind of information. A context can be a larger aggregate of information, a set of simultaneous occurrences that form a sign that is received in its complexity by the interpreting system.

In Bruni (2003; 2007) I have suggested that these processes are mediated by codes that are formed at different hierarchical levels out of an indefinite number of dyadic causal relations, specific "lock and key" interactions, that by their simultaneous occurrence give rise to emergent and "de-emergent" specificities and triadic relations. In the emergence of these semiotic networks "information as specificity" plays a central role together with the interplay between digital and analogical codes in nature.

Experimental settings and field conditions: the importance of the context

One of the major criticisms raised of plant-to-plant communication research is that ubiquitous cues cannot be meaningful information in interactions between damaged and undamaged plants (Dicke, Bruin 2001: 982). This criticism gives over importance to the material medium of information and ignores the triadic nature of information and the specificities created by the context. This is the same problem that arises when dealing with any apparently ubiquitous signals in inter- and intra-cellular communication, like e.g., Ca^{2+}

and cAMP second messengers. And yet, nobody would deny that they convey "meaningful information". What is at stake here is how the emergent interpretant can enact a sort of "categorical sensing" (Bruni 2007; 2008) to recognize the information from the contextual background, a problem that can also be encountered in signal transduction networks.

Knocking out components to evaluate their importance in the working of the system has of course been an extremely useful strategy. This has been widely used from genetic systems (knocking genes) to multitrophic systems (incapacitating or removing one of the participating species). But one of the problems with this strategy is that most knock-outs (genes or species) can be only achieved under controlled conditions. This means that all the contextual parameters of the field cannot be correlated with the knocked component. There can be many factors under natural conditions that affect or are affected by the knocked components. This is a common source of discrepancy between experimental settings and natural conditions. A myriad of simultaneous consensus factors and cues determine the semiotic background in which the component acts. For example, Baldwin *et al.* (2001: 353) report a study on the transcriptional reorganization induced in the plant *Nicotiana attenuata* when it is attacked by its specialist herbivore *Manduca sexta*. The study estimated that more than 500 genes respond to herbivore attack. These coordinated changes parallel the metabolic reconfiguration following pathogen attack and according to the study point to the existence of central herbivore-activated regulators of metabolism. This is one way of relying on patterns in order to "ratchet the system up to a higher informational level". In this case the procedure is not knocking out a gene but a putative inducing factor of a huge set of genes. Even that pleiotropic inducer may be only one detail of a whole context that may potentially change the whole array profile.

Many authors report their worries about discrepancies and inconsistencies between experiments under controlled conditions and field experiments. This recurrent theme is a hint of the importance of the context in communication processes as it was learned from the quorum sensing experience. Seddon *et al.* (1997) argue that much early *in vitro* work was done with the hope that antagonism would be equally effective in the plant environment. Such expectations were naive and unrealistic and many factors other than the direct interaction between antagonist and pathogen play a role *in vivo*. The host plant, the microclimate of the infection court, other microflora and inhabitants of the phyllosphere, environmental parameters and insults (solar radiation, fungicides, etc.) — all contribute and modify this interaction. It is little wonder that many of these earlier biocontrol attempts failed or were invariably inconsistent (Seddon *et al.* 1997: 8). Hatcher and Ayres (1997) ask whether it is possible that mechanisms effective in the laboratory are not effective in the field, and point out that disagreement between laboratory and field results is all too familiar and there may be many reasons for this. Often the growth conditions of the plant are critical. Thus, for example, amounts of pathogenesis-related (PR) proteins produced by plants are strongly dependent upon growth conditions (Hatcher, Ayres 1997: 141).

Fokkema (1997) laments that exploitation of the beneficial effects of mycorrhizal fungi and antagonistic micro-organisms are handicapped by the often observed inconsistency of the beneficial effects under field conditions. He suggests that for future research it is important to identify the major cause of failures. Adequate representation of beneficial organisms at the proper place and time seems the most crucial factor. He advocates for simulation models that consider the responses of introduced populations to a variety of environmental conditions and this will be helpful in selecting more ecologically competent strains. Moreover, the reliability will be improved when we know the conditions under which our introduced micro-organism will work or not (Fokkema 1997: 94). According to Clay (1997), endophyte infections of grasses resemble other mutualistic plant-microbe symbioses, such as mycorrhizas and root nodules that enhance plant growth. Like those symbioses, the benefit of endophyte infection may vary with environmental conditions in conjunction with the genetic background of host plants (Clay 1997: 157), and of the interacting organisms i.e., the "genome space" (Bellgard *et al.* 1999). This evidences the importance of the context when understood as the complex interplay of both the ecological and genealogical hierarchies.

Resistance, virulence and health: subjective categories in ecosystems

A recurrent theme in ecological studies of biodiversity is the occurrence of resistance and pathogenicity. However, these phenomena cannot be fully grasped in terms of trophic interactions because they imply a process of communication and context interpretation. For this reason studies on the *emergence* of "resistance", and its counterpart "pathogenicity", tend to evidence that complex ecological interactions, in spite of the fact that they are acted upon a web of trophic interactions, are not limited to trophic exchanges, and as some research in the past decade points out, semiotic (informational) interactions also have their importance in ecosystem functioning.

Health, and the lack of it, has to be as old as life itself. This problematic concept, as it is intrinsically related to life, expands throughout the entire biological hierarchy from cells to ecosystems. At all these levels the notion presents numerous problems when defining a healthy system. There are different kinds of resistance constantly emerging at different hierarchical levels within biological complexity. There must be general and common principles that lie behind the different forms of resistance originating in nature. Although it is quite different to speak about pest resistance, stress resistance, invasion resistance or to speak about antibiotic resistance, there might be some common principles in different mechanisms (if no other the production of a substance or a context that furnishes protection against something, and *communication* of this pattern to other organisms).

In the preface for the publication of the 5th Symposium of the Society for General Microbiology "Mechanisms of Microbial Pathogenicity" (1955), Howie and O'Hea (1955: x) begin by warning that discussions on how micro-organisms produce disease are very apt to follow a circular course to platitudinous conclusions. That is certainly not my purpose here. However, I think it is useful to explore the analogies of these notions at the different hierarchical levels and their subjective, or relative, nature. In a hierarchical view, notions such as resistance, virulence and health should be seen as subjective categories in ecosystems which take their meaning depending on which side the observer decides to lean towards. These three terms can be generalized to include notions such as survival, predation, pathogenicity, invasiveness, function, balance, equilibrium, resilience and different kinds of symbiosis and interactions (mutualisms, amensalisms, parasitisms, comensalisms, antagonisms and protagonisms) etc. Kratochwil (1999) makes such a generalization when he points out that:

> among themselves, species create bi- and polysystems and thus form so-called bicoenotic links. These interactions between the organisms induce the emergence of characteristics which may contribute to stabilizing the system (quasi-stability in the species composition). Such interaction patterns can be divided into probioses (mutualism, symbiosis, commensalism) and antibiosis (predation, parasitism etc.). (Kratochwil 1999: 13)

Let us consider the paradigmatic case of human bacterial pathogens. The host system senses the presence of the bacterium and resists it through the immune response. The host cell defenses become virulent to the incoming bacterium which deploys resistance to that response through its colonization traits that allow it to avoid, circumvent or subvert the "virulence" of phagocytic cells. Now the resisting bacterium deploys its virulence factors to which the human hosts resists by using antibiotics becoming again virulent to the bacterium that at this point may in turn develop resistance to this new type of human virulence (antibiotics). I am well aware of the specific meaning of the word "antibiotic" and of the need to agree on that definition for practical reasons. However, I will consider Kratochwil's enlarged notion in order to trace some analogies at the different hierarchical levels. The notion of antibiotic implies something that acts against life, something that kills. It is also agreed that it has "natural origin" i.e., it is biosynthetic (a machine gun could also be considered an antibiotic but I will not go that far in my generalization). The standard recognized antibiotics used in medicine are a sort of virulence factor, but they are also a resistance factor seen from the other side of the "fight". They can also be neither of these and simply be signals in a semiotic process. But taken in the most common meaning antibiotics are organisms' chemical weapons. In this sense all types of venoms and poisonous substances produced by organisms should enter into the antibiotic category. Where there is venom there is also resistance, an antidote.

Our cultural dualistic tradition makes it difficult for us to abstract from the warfare vision of the "struggle for life" in order to see the — equally real — other side of the coin consisting of equilibrium, balance and mutualism. It has often been suggested that there has been a coevolutionary "arms race" between, for example, plants and herbivores, as new chemicals are produced and subsequently overcome by insects (Gange, Bower 1997: 116). It may be simplistic to picture the complexity of multitrophic dynamics exclusively with the warfare (arms race) metaphor since it is clear that not all, maybe not even most, relations are antagonistic in nature. Antagonism becomes a subjective category and what may seem a "pathologic" attack at a certain level may turn out to be a healthy mechanism when seen in a larger gestalt. What is "resistance" for one individual or species may signify virulence for another and in turn, the resistance to virulence may be considered an emergent virulence. It may very well be that the escalation process of virulence being resisted and resistance being overcome is at the very base of an unhealthy system (the type of positive feedback that Bateson (1972) called "schismogenesis"). Moreover, virulence and pathogenicity are context-dependent. A virulence factor and the pathogenic organism carrying it may not be such if not in a specific context. That context is a semiotic niche full of signs, some of which trigger virulence out of an otherwise "neutral" organism. For example, in order to invade host cells, *Salmonella* has to simultaneously sense proper levels of oxygen, pH, osmolarity, and an appropriate signal to the PhoP/Q regulon (most probably among other things). If even one of these conditions is unfavorable, the expression of the invasion genes is repressed and *Salmonella* do not invade the host (Falkow 1997: 362). Each of these parameters becomes significant at a specific threshold value. In a different example, Alford and Richards (1999) discuss the local causes to the global decline and losses of amphibian populations, which include ultraviolet radiation, predation, habitat modification, environmental acidity and toxicants, diseases, changes in climate or weather patterns, and interactions among these factors. Many disease agents are present in healthy animals, and disease occurs when immune systems are compromised. They report that declines in populations of *Bufo boreas* between 1974 and 1982 were associated with *Aeromonas hydrophila* infection, and it was suggested that environmental factor(s) (UV-B exposure, changes in pH, pesticides, pollutants etc.) cause sublethal stress in these populations, directly or indirectly suppressing their immune systems. Also a pathogenic fungus largely responsible for egg mortality in one population of *Bufo boreas* in Oregon may have been more virulent to embryos under environmental stress (Alford, Richards 1999: 140).

We could say that there are no pathogens but pathogenic circumstances. In a sense it is the context that becomes pathogenic and at the same time it becomes ill. Since the context (the specific semiosphere) is constantly changing, so is the semiotic niche of a particular system. Any assessment on the emergence of pathogenicity (or conversely the emergence of resistance) has to consider carefully the evolution of the context in relation to the semiotic niche of the potential pathogen, and in relation to its umwelt. The material support of information (e.g., DNA, infochemicals, regulatory elements, etc.) has to be evaluated in the background in which it is inserted, be that genetic, metabolic or ecosystemic. The significance of non-trophic interactions in processes that lead to resistance, virulence and health, when seen as subjective categories in a wider gestalt of co-evolution and symbiosis, means that pathogenesis is basically a semiotic process. The wider gestalt may hint to the existence of a sort of "hierarchical health" in ecosystems.

References

Alford, Ross A.; Richards, Stephen J. 1999. Global amphibian declines: A problem in applied ecology. *Annual Review of Ecology and Systematics* 30: 133–165.

Baldwin, Ian Thomas; Halitschke, Rayko; Kessler, Andre; Schittko, Ursula 2001. Merging molecular and ecological approaches in plant-insect interactions. *Current Opinion in Plant Biology* 4(4): 351–358.

Barea, José-Miguel; Azcon-Agilar, Concepción; Azcon, Rosario 1997. Interactions between mycorrhizal fungi and rhizosphere microorganisms within the context of sustainable soil-plant systems. In: Gange, Brown 1997: 65–77.

Bateson, Gregory 1972. *Steps to an Ecology of Mind*. New York: Chandler Publishing Company.

Bellgard, Matthew I.; Itoh, Takeshi; Watanabe, Hidemi; Imanishi, Tadashi; Takashi, Gojobori 1999. Dynamic evolution of genomes and the concept of genome space. In: Caporale, Lynn Helena (ed.), *Molecular Strategies in Biological Evolution. Annals of the New York Academy of Sciences* 870: 293–300.

Bruin, Jan; Dicke, Marcel 2001. Chemical information transfer between wounded and unwounded plants: Backing up the future. *Biochemical Systematics and Ecology* 29: 1103–1113.

Bruni, Luis E. 2002. Does 'quorum sensing' imply a new type of biological information? *Sign Systems Studies* 30(1): 221–243.

Bruni, Luis E. 2003. *A Sign-theoretic Approach to Biotechnology*. Ph.D. Dissertation, Institute of Molecular Biology, University of Copenhagen.

Bruni, Luis E. 2007. Cellular semiotics and signal transduction. In: Barbieri, Marcello (ed.), *Introduction to Biosemiotics: The New Biological Synthesis*. Berlin: Springer, 365–407.

Bruni, Luis E. 2008. Gregory Bateson's relevance to current molecular biology. In: Hoffmeyer, Jesper (ed.), *A Legacy for Living Systems: Gregory Bateson as Precursor to Biosemiotics*. (Biosemiotics, vol. 2.) Berlin: Springer, 93–119.

Clay, Keith 1997. Fungal endophytes, herbivores and the structure of grassland communities. In: Gange, Brown 1997: 151–169.
Dicke, Marcel; Bruin, Jan 2001. Chemical information transfer between plants: Back to the future. *Biochemical Systematics and Ecology* 29: 981–994.
Emmeche, Claus; Hoffmeyer, Jesper 1991. From language to nature: The semiotic metaphor in biology. *Semiotica* 84(1/2): 1–42.
Eberl, Leo; Winson, Michael K.; Syernberg, Claus; Stewart, Gordon S. A. B.; Christiansen, Gunna; Chhabra, Siri Ram; Bycroft, Barrie; Williams, Paul; Molin, Søren; Givskov, Michael 1996. Involvement of N-acyl-L-homoserine lactone autoinducers in controlling the multicellular behaviour of *Serratia liquifaciens*. *Molecular Microbiology* 20(1): 127–136.
Eberl, Leo; Molin, Søren; Givskov, Michael 1999. Surface Motility of *Serratia liquefaciens* MG1. *Journal of Bacteriology* 181(6): 1703–1712.
Faeth, Stanley H.; Wilson, Dennis 1997. Induced responses in trees: Mediators of interactions among macro- and micro-herbivores? In: Gange, Brown 1997: 201–215.
Falkow, Stanley 1997. What is a pathogen? *ASM News* 63(7): 359–365.
Fokkema, Nyckle J. 1997. Concluding remarks. In: Gange, Brown 1997: 91–95.
Gange, Allan C.; Bower, Erica 1997. Interactions between insects and mycorrhizal fungi. In: Gange, Brown 1997: 115–132.
Gange, Allan C.; Brown, Valerie K. (eds.) 1997. *Multitrophic Interactions in Terrestrial Systems*. (*36th Symposium of The British Ecological Society*.) Oxford: Blackwell Science.
Haber, Wolfang 1999. Conservation of biodiversity — scientific standards and practical realization. In: Kratochwil, Anselm (ed.), *Biodiversity in Ecosystems: Principles and Case Studies of Different Complexity Levels*. The Netherlands: Kluwer Academic Publishers, 175–183.
Hatcher, Paul E.; Ayres, Peter G. 1997. Indirect interactions between insect herbivores and pathogenic fungi on leaves. In: Gange, Brown 1997: 133–149.
Hoffmeyer, Jesper 1997. Biosemiotics: Towards a new synthesis in biology. *European Journal for Semiotic Studies* 9(2): 355–376.
Howie, James William; O'Hea, A. J. 1955. Editor's preface. In: Howie, James W.; O'Hea, A. J. (eds.), *Mechanisms of Microbial Pathogenicity: Fifth Symposium of the Society for General Microbiology*. Cambridge: Cambridge University Press.
Jones, Jonathan D. G. 2001. Putting knowledge of plant disease resistance genes to work. *Current Opinion in Plant Biology* 4: 281–287.
Karban, Richard 1997. Introductory Remarks. In: Gange, Brown 1997: 199–200.
Keel, Christoph; Défago, Geneviève 1997. Interactions between beneficial soil bacteria and root pathogens: mechanisms and ecological impact. In: Gange, Brown 1997: 27–46.
Kratochwil, Anselm 1999. Biodiversity in ecosystems: Some principles. In: Kratochwil, Anselm (ed.), *Biodiversity in Ecosystems: Principles and Case Studies*

of Different Complexity Levels. The Netherlands: Kluwer Academic Publishers, 5–38.

Nimchuk, Zachary; Rohmer, Laurence; Chang, Jeff H.; Dangl, Jeffery L. 2001. Knowing the dancer from the dance: R-gene products and their interactions with other proteins from host and pathogen. *Current Opinion in Plant Biology* 4: 288–294.

Ohgushi, Takayuki 2008. Herbivore-induced indirect interaction webs on terrestrial plants: The importance of non-trophic, indirect, and facilitative interactions. *Entomologia Experimentalis et Applicata* 128: 217–229.

Paul, Nigel D.; Hatcher, Paul E.; Taylor, Jane E. 2000. Coping with multiple enemies: an integration of molecular and ecological perspectives. *Trends in Plant Science* 5(5): 220–225.

Putten, Wim H. Van der; Vet, Louise E. M.; Harvey, Jeffrey A.; Wäckers, Felix L. 2001. Linking above- and belowground multitrophic interactions of plants, herbivores, pathogens, and their antagonists. *Trends in Ecology and Evolution* 16(10): 547–554.

Rice, Scott A.; Givskov, Michael; Steinberg, Peter and Kjelleberg Staffan 1999. Bacterial signals and antagonists: The interaction between bacteria and higher organisms. *Journal of Molecular Microbiology and Biotechnology* 1(1): 23–31.

Roncadori, Ronald W. 1997. Interactions between arbuscular mycorrhizas and plant parasitic nematodes in agro-ecosystems. In: Gange, Brown 1997: 101–113.

Salmond, George P. C.; Bycroft, Barrie W.; Stewart, Gordon S. A. B.; Williams, Paul 1995. The bacterial 'enigma': Cracking the code of cell-cell communication. *Molecular Microbiology* 16(4): 615–624.

Seddon, Barrie; Edwards, Simon G.; Markellou, Emilia; Malathrakis, Nikolaos E. 1997. Bacterial antagonists — fungal pathogen interactions on the plant aerial surface. In: Gange, Brown 1997: 5–25.

Staskawicz, Brian; Parniske, Martin 2001. Biotic interactions: Genomic approaches to interactions of plants with pathogens and symbionts. *Current Opinion in Plant Biology* 4: 279–280.

Swift, Simon; Williams, Paul; Stewart, Gordon S. A. B. 1999. N-Acylhomoserine Lactones and Quorum Sensing in Proteobacteria. In: Dunny, Gary M.; Winans, Stephen C. (eds.), *Cell-Cell Signalling in Bacteria*. Washington, D.C.: American Society for Microbiology, 291–313.

Vos, Matthijs; Vet, Louise E.M.; Wäckers, Felix L.; Middelburg, Jack J.; Putten, Wim H. van der; Mooij, Wolf M.; Heip, Carlo H. R.; Donk, Ellen van 2006. Infochemicals structure marine, terrestrial and freshwater food webs: Implications for ecological informatics. *Ecological Informatics* 1: 23–32.

West, Helen M. 1997. Interactions between arbuscular mycorrhizal fungi and foliar pathogens: Consequences for host and pathogen. In: Gange, Brown 1997: 79–89.

Whipps, John M. 1997. Interactions between fungi and plant pathogens in soil and the rhizosphere. In: Gange, Brown 1997: 47–63.

Chapter 9
Structure and Semiosis in Biological Mimicry

Timo Maran

Summary. Biological mimicry can be described as a structure consisting of two senders (a mimic and a model), a receiver, and their communicative interactions. The distinguishing of three participants in mimicry brings along the possibility to explain mimicry from different perspectives as a situation focused on signal-receiver, mimic, model, or human observer. This has been the foundation for many definitions and classifications of mimicry as well as for some semiotic interpretations. The present chapter introduces some possibilities for defining and classifying mimicry in order to map the dynamical relations between the structure and semiosis in biological mimicry. From a semiotic point of view, the most common property of mimicry seems to be the receiver's inclination to make a mistake in recognition. This allows describing mimicry as incorporating a specific type of semiotic entity — ambivalent sign, — which is understood as an oscillation between one and several signs depending on the actual course of interpretation. Proceeding from Jakob von Uexküll's *Theory of Meaning*, mimicry as any other relation between species is umwelt-dependent i.e., it is conditioned by meanings and functions present for an animal. Therefore also mimic and model, as entities that the receiver fails to differentiate, are first entities of meaning in one's umwelt and are not necessarily representatives of some biological species. The Uexküllian approach allows us to analyze various examples of abstract and semiabstract resemblances in nature. Based on some examples, the biological notion of "abstract mimicry" is reinterpreted here as a situation where the object of imitation is an abstract feature with a universal meaning for many different animal receivers.

Introduction

Biological mimicry has become an increasingly popular topic in semiotics in the last decade. Different authors have provided semiotic interpretations of the complex mimicry systems of fireflies (El-Hani *et al.* 2010) and dynamic camouflage of cephalopods (Renoue, Carlier 2006), discussions of mimicry for zoosemiotics (Martinelli 2007: 45–47) and developments in semiotic theory of mimicry (Kleisner, Markoš 2005; Kleisner 2008; Maran 2005; 2007; 2011). The aim of this short survey is twofold: to map the dynamical relations between the structure and semiosis in biological mimicry, and to give an overview of possible interpretations in biological mimicry. By doing so, it is proposed that semiotic accounts of mimicry should develop towards a more systemic understanding of the phenomenon.

Biological mimicry can be described as a structure consisting of three participants: a mimic, a model, and a receiver, and their communicative interactions. From the perspective of communication theory, these three participants can be divided between the position of sender and the position of receiver so that the mimic and the model occupy the position of sender as opposed to that of signal-receiver. This tripartite structure has been the foundation for many definitions and classifications of mimicry. The relations between the three participants commonly pointed out in mimicry definitions are: (1) similarity between colors, signals or species; (2) deception of one participant, or a participant's inability to recognize the difference; (3) some use or benefit for, or increase/decrease of the fitness of the participants. For instance, British entomologist Richard I. Vane-Wright defines mimicry as follows: "Mimicry occurs when an organism or group of organisms (the mimic) simulates signal properties of a second living organism (the model), such that the mimic is able to take some advantage of the regular response of a sensitive signal-receiver (the operator) towards the model, through mistaken identity of the mimic for the model" (Vane-Wright 1976: 50).

The distinguishing of three participants in mimicry and their relationships brings along the possibility to explain mimicry from different perspectives, that is, as a situation perceived by either the signal-receiver, the mimic, the model, or the human observer. In early studies mimicry was regarded predominantly from the viewpoint of human researcher and considered rather as a taxonomic disorder or as a fallacious similarity between different species. For instance in 1862 Henry Walter Bates specifies mimicry to be "resemblances in external appearance, shape, and colors between members of widely distinct families [...] The resemblance is so close, that it is only after long practice that the true can be distinguished from the counterfeit, when on the wing in their native forests" (Bates 1862: 502, 504). Later, other perspectives became more eminent. Studies of warning coloration introduced the view of mimicry as a parasitic phenomenon that takes advantage of and at the same time is dependent on normal communication (e.g., Cott 1957: 396–397). Classical studies on mimicry by Jane Van Zandt Brower and Lincoln Pierson Brower launched the understanding of resemblance between mimics and models as a behavioral dilemma for the signal-receiver (Brower 1960; Brower, Brower 1962).

Semiotic interpretations of mimicry: from simulation to ambivalent sign

In semiotics, the specific aspects that have been emphasized when discussing biological mimicry also seem to depend largely on the researcher's position with regard to the triad of mimic, model, and signal-receiver. For instance Thomas Sebeok tends to emphasize the position of mimic, when describing mimicry as an example of iconicity in nature (Sebeok 1990: 95–96). From the position of mimic, the process of changing itself or the surrounding environment in order to resemble the model can be considered as a creation of iconic resemblance. This preference is well illustrated by Sebeok's description of the behavior of the Asiatic spider who changes "its surroundings to fit its own image by fabricating a number of dummy copies of itself to misdirect predators away from its body, the live model, to one of several replicas it constructs for that very purpose" (Sebeok 1989: 116).

Such an approach is in compliance with Sebeok's later theoretical stand: to describe different types of sign in connection with various modeling strategies, i.e., rather from the position of the utterer and sign creation than that of the receiver and sign perception. For instance, in the book *The Forms of Meaning: Modeling Systems Theory and Semiotic Analysis*, iconic signs are defined on the basis of the features of sign creation: "A sign is said to be iconic when the modeling process employed in its creation involves some form of simulation. Iconic modeling produces singularized forms that display a perceptible resemblance between the signifier and its signified. In other words, an icon is a sign that is made to resemble its referents in some way" (Sebeok, Danesi 2000: 24). A rather similar approach to mimicry is provided by the hermeneutical Portmannian interpretation, where mimetic resemblances are considered as part of self-expression of an individual (e.g., Kleisner, Markoš 2005).

An alternative possibility to analyze mimicry as a semiotic phenomenon is to focus on the position of signal-receiver. A mimicry situation may appear to the receiver very differently from how it appears to the sender. This change is first rooted in a common feature of communication: the emergence of shifts in meanings due to the asymmetry of the processes of formulating and interpreting, coding and decoding. Theatre semiotician Tadeusz Kowzan has described this as different aspects of sign, which are expressed in the different phases of communication. For instance a sign can be mimetic in its creation and iconic in its interpretation (Kowzan 1992: 71). In mimicry, however, the difference of meaning for the sender and the receiver seems to be a more fundamental property. Alexei A. Sharov has explicated mimicry with the term "inverse sign", where sign has a positive value for the sender ("transmitter" in his terminology), but negative for the receiver. Sharov describes female fireflies, which imitate light signals of other species to attract their males in order to eat them, as an example of such inverse signs. Sharov specifies that "an inverse sign is always an imitation of some other sign with positive value for the receiver" (Sharov 1992: 365).

Similarly to many other cases of animal semiosis, in mimicry for the signal-receiver the sign relation is formed from the search image or perceived features of an organism (representamen), the organism as it is physically capable of being interacted with (object), and the meaning connected with the applicability of the organism (interpretant). The common denominator of mimicry seems to be the signal-receiver's effort to make the correct recognition in a situation where perceptibly similar objects or organisms may be present (see Maran 2001: 332–334). The difference between the model and the mimic for the signal-receiver may be manifested for instance in the following oppositions: discernible object *versus* perceptual noise, eatable *versus* uneatable item, safe *versus* dangerous organism. The oppositions often go together with diametrically opposite aspirations to react (e.g., catch *versus* flee). The differentiation of mimics from models depends on many contextual factors (such as the physiological status of the participants, or the specific location of the mimicry situation) and therefore it reappears in each and every act of communication. Because of this it is not possible to conclude whether there are one or two signs or sign complexes involved in mimicry.

In our effort to deal with ambiguous sign complexes, the American semiotician Charles Morris can offer us some guidance. In "Signs, language, and behavior" Morris introduces the term "sign family", defining it as a group of signs, which have the same meaning for the interpreter: "A set of similar sign-vehicles which for a given interpreter have the same significata will be called a sign-family" (Morris 1971: 96). In accordance with his behaviorist stand, Morris unites signs into a sign family on the basis of a similar behavioral reaction released by the interpreter. In connection with the concept of sign family Charles Morris also points out that a sign may, but need not, have only one meaning. He contrasts unambiguous and ambiguous signs: "A sign-vehicle is *unambiguous* when it has only one significatum (that is, belongs to only one sign-family); otherwise ambiguous" (Morris 1971: 97).

The concept of ambiguous signs seems to cover different types of relations between meanings. First, there can be situations where meanings complement each other, and second, there can be situations when different interpretations or meanings are in opposition and exclude each other. For mimicry, the second type of ambiguousness is more characteristic. In its *umwelt* the interpreter cannot combine interpretations that correspond to the mimic and the model species but needs to choose between these. Therefore it would be more correct to call such sign combination *ambivalent sign* instead of ambiguous sign. *Ambivalent sign can be described as a sign structure, which fluctuates between one and two signs and where the actual composition and number of signs emerges in the course of interpretation.* Such ambivalence has structural importance in mimicry. The perceptual similarity of mimics and models, and the opposition in meanings are components of evolutionary conflict between the mimic and the signal-receiver and an important feature of the communicative regulation between them.

Uexküllian perspective on mimicry

Besides analyzing meanings that different objects obtain in umwelten of various organisms, biosemiotic research can also focus on diverse relations between animals to discover meaningfulness there. In *Bedeutungslehre,* Jakob von Uexküll describes correspondences between body plans and umwelten of different animals as counterpoints of meaning. The different umwelten are mediated by functional cycles, where animals obtain the positions as meaning utilizers and meaning carriers for each other through the perceptual and effectual activity. According to Uexküll these counterpoints of meaning modify entire structures of animal bodies as well as their lifecycles. "The meaning of all plant and animal organs as utilizers of meaning-factors external to them determines their shape and the distribution of their constituent matter" (Uexküll 1982: 37). These meanings can also be mediated by cue-carriers that are distinct from the animal's body, such as the squeaking sound standing for the bat in the moth's umwelt, but also by a completely distinct organism who acts as a meaning carrier. Here Uexküll presents an example of the male bitterling (*Rhodeus*) in which it is not the female fish that causes mating coloring to occur but the sight of the pond-mussel. The bitterlings spawn into the mussel gills where the young fish larvae can later safely grow (Uexküll 1982: 55).

Structures in nature that mediate meanings (i.e., external signs) make it possible to consider mimicry in the Uexküllian framework of contrapuntal correspondences. With respect to mimicry, Uexküll mentions two examples: the angler-fish *Lophius piscatorius*, who uses a long and movable appendage to lure prey fish, and butterflies that carry colorful eye-resembling spots which scare off insectivorous birds. Uexküll sees these examples as an extension of meaning rules that organize forms in nature. The form shaping of the prey is in these cases not directly connected to the form shaping of the predator, but correspondence is achieved due to some other image or shape-schemata present in the animal's umwelt (Uexküll 1982: 58–59).

Uexküll's *Bedeutungslehre* opens a significant aspect of relations between species, which should be considered as the biosemiotic ground for interpreting mimicry. That is, the relations between different species, to the extent that these are based on communication, are umwelt-dependent, i.e., they are conditioned by the meanings and functions present for the animal. Concerning mimicry, the Uexküllian approach means that any deceptive resemblance should be considered first from the viewpoint of the participants' umwelten. This premise brings along some quite significant consequences for the semiotic interpretation of biological mimicry.

First, it means that the common description of mimicry as a resemblance between two species covers only rather limited cases among many possible similarities. As taxonomic classifications of biological species are the product of human culture and thus specific to the human umwelt, the animal receiver may distinguish between perceived organisms completely differently. For instance, the taxonomic diversity of bees, bumblebees and wasps as it is described in biology may in flycatchers' umwelt form just one group of buzzing and colorfully striped flyers who tend to sting upon catching. If this is so, then Müllerian mimicry can be noticed only by a human observer, as from the viewpoint of the signal-receiver there are no distinct classes involved.

The second implication is that for the signal-receiver, neither the mimic nor the model needs to be a whole organism but can be just a part of an organism both in spatial or temporal terms or just a perceptible feature. For example, one can say that in a loose sense, the fly orchid *Ophrys insectifera* as a plant is a mimic, but for solitary wasps who pollinate the flower because of mistaking it for female wasps, the similarity is much more concrete. In the wasp's umwelt only a blossoming plant can be confused with the mate, and only in the right weather conditions when the pheromone-like smell of the fly orchid floats in the air.

The third implication of the Uexküllian view, which emphasizes the role of meanings in the relations and communication between species, would be an understanding that in animal umwelten there may exist intense meanings which need not have any direct or strong relations to any specific physical forms. Instead, an animal itself can attribute such meanings to many different objects that match these meanings. Such universal meanings are for instance "sudden change", "unfamiliarity", "possible danger". Often triggering behavioral responses like halting, fleeing or curiosity to gather more information, these meanings can also become a source for imitation.

Probably the most general level of abstraction is present in warning displays. A well-known defense strategy that uses abstract resemblance is the behavioral adaptation of many reptiles and amphibians to make themselves appear larger in the presence of danger. For instance, upon noticing a snake, the common toad *Bufo bufo* lifts itself up from the ground and emits strange growling sounds while at the same time its body becomes bloated because of the inhaled air. The aim of this behavior is to become more noticeable and thus to convince the snake that this particular toad does not belong to the group of prey animals. From the viewpoint of mimicry studies, we can say that the toad is mimicking in the most abstract manner the sign complex (model) which in the snake's umwelt corresponds to animals that are too big to catch.

French zoologist Georges Pasteur excludes such examples from his profound species-based mimicry classification on the grounds that "the model is not an actual species" and describes these under the terms "semi-abstract and abstract homotypy" (Pasteur 1982: 191). As I indicated before, the question of resemblance to a model not belonging to any concrete species may actually go deeper than just classificatory issues and pertain to the common biological understanding of relations between species, which focuses on physical forms and properties and largely ignores perceptual features and meanings for the animals themselves (for discussion, see Maran 2007).

The explanatory power of the biosemiotic view becomes apparent in analyzing the cases of abstract mimicry where the similarity between the mimic and model species is approximate or diffuse. An example of such "imperfect mimicry" (Sherratt 2002) characteristic of Holarctic is the combination of yellow–black warning coloration of many *Hymenoptera* (wasps, bees, bumblebees) and their imitations on many levels of exactness by hover-flies *Syrphidae*, but also by some moths, beetles, dragon-flies and other insects. Most biological approaches regard imperfect mimicry as some deviation from the "normal situation" of drive toward absolute similarity. Such approaches seek to explain imperfectness with specific environmental conditions or ecological relations or try to find some other factor that would compensate for deviation (Gilbert 2005).

From the biosemiotic perspective we can make a principally different suggestion: hover-flies do not imitate any concrete species, but rather a certain combination of colors, which have the meaning of unpalatability or danger for a large group of animal receivers. In other words, the attention of the signal-receiver is focused on the relations between the insect and the conspicuous color pattern with its possible meaning, not on comparing different insects and typifying these. In such case it is not the exact resemblance that becomes decisive, but whether they expose their common color pattern recognizably enough and whether the signal-receiver is familiar with the meaning of the pattern.

Conclusions

In the present chapter several communicational and semiotic aspects of biological mimicry have been discussed. The tripartite nature of mimicry systems makes it possible to characterize mimicry from various viewpoints. Focusing on the position of mimic the mimicry could be understood as the example of iconicity in nature, as Thomas Sebeok does it. Focusing on the position of signal-receiver, mimicry becomes a dilemma of recognition. From the Uexküllian perspective of meaningful relations in nature, mimicry becomes the mediated correspondence between umwelten. In order to describe mimicry as an integrated, whole semiotic structure, these different perspectives must be taken into account.

References

Bates, Henry W. 1862. Contributions to an insect fauna of the Amazon valley, *Lepidoptera: Heliconidæ*. *Transactions of the Linnean Society of London* 23: 495–566.

Brower, Jane Van Zandt 1960. Experimental studies of mimicry. IV. The reactions of starlings to different proportions of models and mimics. *American Naturalist* 94: 271–282.

Brower, Jane Van Zandt; Brower, Lincoln Pierson 1962. Experimental studies of mimicry. 6. The reaction of toads (*Bufo terrestris*) to honeybees (*Apis mellifera*) and their dronefly mimics (*Eristalis vinetorum*). *The American Naturalist* 96: 297–308.

Cott, Hugh B. 1957. *Adaptive Coloration in Animals*. London: Methuen.
El-Hani, Charbel N.; Queiroz, João; Stjernfelt, Frederik 2010. Firefly femmes fatales: A case study in the semiotics of deception. *Biosemiotics* 3(1): 33–55.
Gilbert, Francis 2005. The evolution of imperfect mimicry in hoverflies. In: Fellowes, Mark D. E.; Holloway, Graham J.; Rolff, Jens (eds.), *Insect Evolutionary Biology*. Wallingford: CABI Publishing, 231–288.
Kleisner, Karel 2008. Homosemiosis, mimicry and superficial similarity: Notes on the conceptualization of independent emergence of similarity in biology. *Theory in Biosciences* 127(1): 15–21.
Kleisner, Karel; Markoš, Anton 2005. Semetic rings: Towards the new concept of mimetic resemblances. *Theory in Biosciences* 123(3): 209–222.
Kowzan, Tadeusz 1992. *Sémiologie du Théâtre*. Paris: Nathan.
Maran, Timo 2001. Mimicry: Towards a semiotic understanding of nature. *Sign Systems Studies* 29(1): 325–339.
Maran, Timo 2005. *Mimikri kui kommunikatsioonisemiootiline fenomen*. [Mimicry as a communication semiotic phenomenon]. *Dissertationes Semioticae Universitatis Tartuensis* 7. Tartu: Tartu University Press.
Maran, Timo 2007. Semiotic interpretations of biological mimicry. *Semiotica* 167(1/4): 223–248.
Maran, Timo 2011. Becoming a sign: The mimic's activity in biological mimicry. *Biosemiotics*, Forthcoming DOI: 10.1007/s12304-010-9095-8.
Martinelli, Dario 2007. *Zoosemiotics: Proposal for a Handbook*. (Acta Semiotica Fennica 26.) Imatra: International Semiotics Institute at Imatra.
Morris, Charles 1971. Signs, language, and behavior. In: Morris, Charles, *Writings on the General Theory of Signs*. The Hague: Mouton, 73–397.
Pasteur, Georges 1982. A classificatory review of mimicry systems. *Annual Review of Ecology and Systematics* 13: 169–199.
Renoue, Marie; Carlier, Pascal 2006. Au sujet des couleurs de céphalopodes — rencontre de points de vue sémiotique et éthologique. *Semiotica* 160(1/4): 115–139.
Sebeok, Thomas A. 1989. Iconicity. In: Sebeok, Thomas A., *The Sign & Its Masters*. Lanham: University Press of America, 107–127.
Sebeok, Thomas A. 1990. Can Animals Lie? In: Sebeok, Thomas A., *Essays in Zoosemiotics*. (Monograph Series of the Toronto Semiotic Circle 5.) Toronto: Toronto Semiotic Circle; Victoria College in the University of Toronto, 93–97.
Sebeok, Thomas A.; Danesi, Marcel 2000. *The Forms of Meaning: Modeling Systems Theory and Semiotic Analysis*. Berlin: Mouton de Gruyter.
Sharov, Alexei A. 1992. Biosemiotics: A functional-evolutionary approach to the analysis of the sense of information. In: Sebeok, Thomas A.; Umiker-Sebeok, Jean; Young, Evan P. (eds.), *Biosemiotics: The Semiotic Web 1991*. Berlin: Mouton de Gruyter, 345–373.

Sherratt, Thomas N. 2002. The evolution of imperfect mimicry. *Behavioral Ecology* 13(6): 821–826.
Uexküll, Jakob von 1982. Theory of meaning. *Semiotica* 42(1): 25–82.
Vane-Wright, Richard I. 1976. A unified classification of mimetic resemblances. *Biological Journal of the Linnean Society* 8: 25–56.

Chapter 10
Semiosphere Is the Relational Biosphere

Kaie Kotov and Kalevi Kull

Summary. The concept of semiosphere was first formulated by Juri Lotman in 1982. He compared it with biosphere, the concept as described by Vladimir Vernadsky. In concordance with Sebeok's thesis on coextensiveness of life and semiosis, Jesper Hoffmeyer has introduced the concept of semiosphere as covering semiosis of all life processes. According to Vernadsky, biosphere is the *matter* that is chemically changed in result of life processes (whereas noosphere is the matter that is chemically changed in result of human mind). Semiosphere is not the matter but the whole set of semiosic *relations*. We describe the main elements of the model of semiosphere, as introduced by Lotman. Theory of semiosphere can be seen as a basis for general semiotics.

Semiosphere

The concept of semiosphere was first put forward by Juri Lotman in the context of cultural semiotics. He introduced the term in the article "On semiosphere" (Lotman 1984[1]) and elaborated it further in *Universe of the Mind* (Lotman 1990) and *Culture and Explosion* (Lotman

---

[1]See the English translations of Lotman's (1984) article in Lotman (1989; 2005); cf. Lotman (2010: 201ff). The first conference in which Juri Lotman spoke about the concept of semiosphere was seemingly the *8th Estonian Spring School in Theoretical Biology* in May 1982 (see Kull 2006).

1992[2]). The fundamental insight that underlies Lotman's concept is the notion that the smallest functioning mechanism is not a single sign or single text or single semiotic system, but the whole semiotic space whose internal organization is created and maintained by the multiple sign processes occurring at the different levels of the multifaceted and multilevelled system of communication.

Semiosphere is a semiotic space that is necessary for the existence and functioning of languages and other sign systems. All semiotic systems are "immersed" in a semiotic space and "can only function by interaction with that space" (Lotman 1990: 125). A sign cannot make sense except in the context of other signs. In other words, semiosphere is a sphere of semiosis and *an experience thereof*; and as such, it is a prerequisite for any single act of communication to be interpreted as one. Lotman's insistence on the prior existence of semiotic space in relation to single texts as well as to the interrelationship of texts within semiosphere has an analogue in C. S. Peirce's principle *omne symbolum e symbolo*. In biology, an analogical relationship has been formulated as Francesco Redi's law (in the 17th century): *omne vivum e vivo*. Another version of it is Jakob von Uexküll's (1920: 6) *every design is from design*.[3]

Some fundamental assumptions about the structural properties of the semiotic space of culture have already been formulated in *Theses on the Semiotic Study of Cultures*:

> In the study of culture the initial premise is that all human activity concerned with the processing, exchange, and storage of information possesses a certain unity. Individual sign systems, though they presuppose immanently organized structures, function only in unity, supported by one another. None of the sign systems possesses a mechanism which

---

[2]English translation — Lotman (2009).

[3]We could add here Gregory Bateson's observation: "The mental world is only maps of maps, ad infinitum" (Bateson 2000 [1972]: 460).

would enable it to function culturally in isolation. (Ivanov
et al. 1998 [1973]: 33)

These statements along with the insistence on the fundamental duality between verbal (discrete) and pictorial (iconic) modelling systems (i.e., "code duality")[4] form the basis of Lotman's writings throughout his scholarly career. Yet the concept of semiosphere marks a shift from strictly structuralist concerns to addressing more explicitly the dynamics and organicism (cf. Mandelker 1994; 1995) of cultural systems.

The concept of semiosphere draws on an analogy with Vladimir I. Vernadsky's concept of biosphere. In his letter to Boris Uspensky, written in 1982, Lotman first outlines his position:

> While reading Vernadsky, I was stunned by one of his statements. You know [...] my opinion that a text can exist (i.e., is socially acknowledged as a text) if another text precedes it, and that any advanced culture must have been preceded by an advanced culture. And now I have discovered Vernadsky's thought, deeply founded on experience of exploring cosmic geology, that life can originate only from the living, i.e., only if it is preceded by it. [...] Only the precedement of semiotic sphere makes message out of message. Only the existence of intellect explains the existence of intellect. (Lotman 1997: 630)

J. Lotman interprets Vernadsky's concept of biosphere as the totality and the organic whole of living organisms and life processes. Accordingly, he proposes that semiosphere is to be understood as the totality of texts, languages, and sign processes that are interconnectedly bound to each other. Mihhail Lotman has pointed out that this assumption has a necessary methodological implication: "In the field of culture simple models do not precede the complicated ones, but, vice versa, simple models are the result of investigator's

---

[4] Cf. Hoffmeyer, Emmeche 1991.

abstraction or the result of reduction or degeneration of complicated systems" (M. Lotman 2001: 98; see also M. Lotman 2002). The concept of semiosphere as put forward by J. Lotman forms the core of the holistic theory of culture that takes its departure from the assumption that "the unit of semiosis, the smallest functioning mechanism is not a separate language, but the whole semiotic space of culture in question" (Lotman 1990: 125). *The elementary unit of semiosis is semiosphere*.[5]

Functionally, semiosphere can be characterized as a "thinking" system that is able to: (1) transmit available information; (2) to create new information that is not simply deducible according to a set of algorithms from already existing information, but which is to some degree unpredictable; (3) to preserve and reproduce information, that is, any semiotic system has its own memory. Semiosis in this system can be understood mainly in terms of translation and dialogue. "Translation is a primary mechanism of consciousness. To express something in another language is a way of understanding it" (Lotman 1990: 127). The need for dialogue, the dialogic situation precedes both the real dialogue as well as the existence of a language in which to conduct it: dialogic situation creates the common language that underlies the translation of messages (Lotman 1990: 143).

One of the most important categories in the semiosphere is that of *boundary*. The system is able to engage in dialogic processes only if its structural identity is established. The beginning point for semiotic individuation is the binary distinction of *inside* versus *outside*: the boundary of the semiosphere can be defined as the outer limit of a first person form. Therefore, semiosphere is a "bounded" system in a sense that it is distinguished from and cannot have (a non-translational) contact with non-semiotic or alien semiotic systems. Semiosphere is closed as a system of (all types of) knowing. On the other hand, semiotic boundary is to be conceived as an abstraction, a series of bilingual filters or membranes that enable the translation

---

[5]This should also concern the elementary unit of life (see Kull 2002).

of messages from one semiotic system into another. In this way, boundary that is defined as an at least double-coded system of translation filters both determines the identity of the system and allows the translation of messages between the different semiotic systems.

The category of boundary leads to two further characteristics of semiosphere: *binarism* and *asymmetry*. The translation mechanism responsible for the generation of new meaning in semiosphere presupposes at least two semiotically different participants that are mutually untranslatable. The entire space of the semiosphere is transected by boundaries and "since in the majority of cases the different languages of the semiosphere are semiotically asymmetrical, i.e., they do not have mutual semantic correspondences, then the whole semiosphere can be regarded as a generator of information" (Lotman 1990: 127). Asymmetry is also manifest in the relationship between *centre* and *periphery*, between the defining core that tends toward the *homogeneity* in the semiosphere through the mechanisms of *autocommunication*, and the accelerated semiotic processes (fluctuations) that reveal the internal *heterogeneity* of the semiosphere.

It follows from the above statements that, firstly, the unity of semiosphere is always a more-or-less phenomenon and can be conceived only insofar as we consider the *self-description* of the given system: the semiosphere is marked by heterogeneity. It is filled up with multiple semiotic systems "that relate to each other along the spectrum which runs from complete mutual translatability to just as complete mutual untranslatability" (Lotman 1990: 125). Secondly, semiosphere constantly comes into contact with other semiotic systems that formerly lay beyond its boundaries. Thirdly, the dynamics of the subsystems within the semiosphere is marked by heterogeneity as well. As a conclusion, semiosphere can be described as a "semiotic continuum", a heterogeneous yet bounded space that is in constant interaction with other similar structures. The points of contact between different systems (which, in turn, are part of a heterogeneous space of a higher order) enable the emergence of new meaning, that is, the deviation from the already established codes, models, or habits of the given system.

The concept of semiosphere also reflects the influence of the theory of self-organizing systems proposed by Ilya Prigogine and Isabelle Stengers (1984). For Lotman, Prigogine's ideas have two important implications. First, the stochastic and lawful (chance and necessity) are revealed as the two sides of the same coin. Second, this leads to the separation in time and causality. This enabled Lotman to formulate a non-progressive view on dynamics of culture where the rapid (stochastic, explosive) processes in culture alternate with the periods of normal development. Semiosphere appears thus as a self-organizing system that undergoes constant renewal due to the emergence of new meanings created in the process of internal and external translation. Indeed, the basic mechanism for the creation of new meanings in culture appears to be the existence of a certain amount of mis-understanding or mis-communication that is built into the very structure of its semiotic space while the structural stability of any semiotic system is maintained by the self-descriptions created by the system in the process of autocommunication. The communication between different semiotic systems tends towards the increase in heterogeneity while autocommunication tends towards the increase in homogeneity.[6]

Biosphere

Biosphere is a global network of life, a self-regulating system that embraces both the living organisms (biota) and their abiotic environment to the extent it is involved in the processes of life, including the troposphere, the ocean, and the upper envelopes of the earth crust.

---

[6]See also the 17 different definitions of semiosphere (collected in Kull 2005) which describe and reflect some further features of semiosphere. Analysis of some aspects of the concept of semiosphere can be found in Alexandrov (2000); Chang (2003); Kawade (1998); Kotov (2002); Machado (2007); Markoš (2004); Nöth (2006); Torop (2005).

Living matter, dispersed in the form of living organisms and sharply separated from its inert environment, is the most important source of self-regulation and stability in the biosphere.

The foundational treatises on biosphere (and a conception related to it, noosphere) are attributed to Pierre Teilhard de Chardin and Vladimir I. Vernadsky. However, the modern usage of the term "biosphere" began with Eduard Suess, a palaeontologist and geologist at the University of Vienna, who adopted the term in 1875. His brief account is ambivalent and can be interpreted as referring either to (1) the sum total of living organisms or (2) a geosphere created and organized by the life processes. The former view was taken up by Teilhard de Chardin while Vernadsky combines the two meanings to account for the biosphere as a geological envelope and a self-organizing system.

For Vernadsky, the biosphere is defined by the "cyclical processes of atom exchange between living matter and inert matter in the biosphere" (Vernadsky 1977: 111), i.e., the biogenic migration of atoms. The processes of atom migration are also a fundamental source of change in the biosphere. According to the principles of biogeochemistry formulated by Vernadsky, the evolution in the biosphere is an irreversible process that proceeds "in the direction of increasing the level of self-regulation and stability" (Levit 2001: 61). One of the basic methods to achieve this is "to increase the intensity and the complexity of biogenic migration of atoms" (Levit 2001: 65), i.e., the basic determinant in the evolution of the biosphere is the growth of the atom exchange caused by the life processes:

> The biosphere appears in biogeochemistry as a peculiar envelope of the Earth clearly distinct from the other envelopes of our planet. The biosphere consists of some concentric contiguous formations surrounding the whole Earth and called geospheres. The biosphere has possessed this perfectly definite structure for billions of years. This structure is tied up with the active participation of life, is conditioned by

life to a significant degree and is primarily characterised by dynamically mobile, stable, geologically durable equilibria which, in distinction from mechanical structures are quantitatively fluctuating within certain limits in relation to both space and time. (Vernadsky 1977: 120)

One of Vernadsky's central assumptions is that living and inert natural bodies are connected only by a biogenic flow of atoms: living matter in biosphere is embedded in its inert environment, yet it is clearly distinct from it from both structural and energetic points of view so far as to say that it constitutes an independent space–time that functions according to the laws of its own. There are almost no transitional forms between living and inert matter. Vernadsky draws here on the observations made by Louis Pasteur and Pierre Curie about the molecular dissymmetry discerning living organisms from inert matter: for the dissymmetry to occur, it presupposes a space whose organization is also dissymmetrical. In other words: dissymmetrical effects can be brought about only by a dissymmetrical cause. This assumption leads to one of the most important statements in Vernadsky's theory: the impossibility of abiogenesis in biosphere; life presupposes life but also that life appears from the very beginning in the form of the biosphere. According to Vernadsky, then, biosphere is what we might call an emergent phenomenon.

Since 1968, James Lovelock, without having read Vernadsky, or at least so it seems, began to develop the so-called Gaia hypothesis, which states that the physical conditions of the Earth's surface, oceans, and atmosphere have been made fit for life and are maintained in this state by the living organisms themselves. According to a definition proposed by Lovelock, Gaia is:

a complex entity involving the Earth's biosphere, atmosphere, oceans, and soil; the totality constituting a feedback or cybernetic system which seeks an optimal physical and chemical environment for life on this planet. The maintenance of relatively constant conditions by active control may be conveniently described by the term "homeostasis". (Lovelock 1979: 10)

In this respect, Vernadsky's concept of biosphere as a self-regulating system anticipates and already includes many crucial claims put forward in Lovelock's concept.

Noosphere

According to Vernadsky, by the beginning of the 20th century, biosphere had reached in its evolution a transitional period from biosphere to noosphere. In this stage, the central stabilizing force would not be life but human thought; more precisely, scientific thought. In this respect, the latter is a function of the biosphere and thus a geological phenomenon. Therefore, in noosphere, the functions of the biogenic energy created by life process would be taken over by "the energy of human culture" — a term coined by Vernadsky to denote the transformative force created by the activity of the human mind.

The rise of civilization is a geological necessity, its continuous development is related to the dissymmetry of time in life process whose function is scientific thought; according to Vernadsky:

> a civilization of "cultural humanity" (being a form of a new geological force created in the biosphere) *cannot disappear or cease to exist*, for it is a great natural phenomenon corresponding historically, or more correctly, geologically, to the established organization of the biosphere. Forming the noosphere, the civilization becomes connected through all its terrestrial roots to its terrestrial envelope (*biosphere*), which has never happened in the previous history of mankind to a comparable degree. (Vernadsky 1977: 33; English quotation in Levit 2001: 77)

Noosphere, therefore, is not a layer *in* the biosphere but it *is* the biosphere, where the central role belongs to the "energy of human culture" (Vernadsky 1977: 95), to "scientific thought".

Teilhard de Chardin considered biosphere as an aggregate of terrestrial living organisms. Accordingly, Teilhard's noosphere is a "thinking layer" (Teilhard 1967: 202), one more envelope around and over the biosphere, its appearance marking not the next stage in the evolution of the biosphere but the rise of the split between the intelligence and its material matrix leading to the death of the Earth. Therefore, noosphere is only a transitional stage in the further development of supreme consciousness, "the end of all life on our globe, the death of the planet, the ultimate phase of the phenomenon of man" where "mankind, *taken as a whole*, will be obliged [...] to reflect upon itself at a single point" (Teilhard 1967: 300, 315). Little by little, the human minds extend:

> the radius of their influence upon this earth which, by the same token, shrank steadily. [...] Through the discovery yesterday of the railway, the motor car and the aeroplane, the physical influence of each man, formerly restricted to a few miles, now extends to hundreds of leagues or more. Better still: thanks to the prodigious biological event represented by the discovery of electro-magnetic waves, each individual finds himself henceforth (actively and passively) simultaneously present, over land and sea, in every corner of the earth. (Teilhard de Chardin 1967: 240)

This paragraph from Teilhard de Chardin was later recalled by Marshall McLuhan in a passage discussing the "new electronic interdependence" that "recreates the world in the image of a global village": "This externalization of our senses creates what de Chardin calls the 'noosphere' or a technological brain of the world" (McLuhan 1962: 31–32).

One must make a clear distinction between semiosphere and noosphere: "The existence of noosphere is material and spatial, it encompasses a part of our planet, whereas the space of semiosphere is of an abstract kind" (Lotman 1984: 6). However, as noted by Vjačeslav V. Ivanov, the noosphere is unthinkable without the semiosphere (Ivanov 1998: 792). Noosphere, acoording to Vernadsky, is the *matter* that is chemically changed in result of human mind. Semiosphere is not the matter but the whole set of semiosic *relations*, i.e., the sphere of *semiosis*.

Biosphere as semiosphere

In the semiotic context, Thomas A. Sebeok has defined biosphere as "that parcel of the planet earth which comprises life-signs [...]. The biosphere is where we live and what we are; but, although it is the only domicile we possess thus far, we are neither its sole inhabitants or constitute anything approaching a plurality in — let alone tenured — occupancy" (Sebeok 1997: 113).

In 1993, Danish biosemiotician Jesper Hoffmeyer adopted the term "semiosphere" (independently of Lotman) to designate the totality of semiotic processes going on planet Earth.[7] According to Hoffmeyer, semiotic quality of life manifests itself in tendency to take habits, that is, in the tendency to evolve into new regularities.

> Living systems exhibit extreme semiogenic behaviour through the process I have termed "semethic interaction". Semethic interactions refer to interactions in which regularities (habits) developed by one species (or individual) successively become used (interpreted) as signs by the individuals of the same or another species, thereby eliciting new habits in this species eventually to become — sooner or later — signs for other individuals, and so on in a branching and

---

[7]Marking a difference in scope between Lotman's and Hoffmeyer's concepts of semiosphere, John Deely proposed to use the term *signosphere* "as a term more appropriate for the narrower designation of semiosphere in Lotman's sense" (Deely 2001: 629). However, later he treats these as synonyms (Deely 2010: 79, 157, 158).

unending web integrating the ecosystems of the planet into a global semiosphere. (Hoffmeyer 1998: 287)

Accordingly,

the semiosphere is a sphere just like the atmosphere, the hydrosphere, and the biosphere. It penetrates to every corner of these other spheres, incorporating all forms of communication: sounds, smells, movements, colors, shapes, electrical fields, thermal radiation, waves of all kinds, chemical signals, touching, and so on. In short, signs of life. (Hoffmeyer 1996: vii)

Hoffmeyer proceeds from the semiotic theory of Peirce by considering semiosis central to the understanding of life processes: "signs, not molecules are the basic units in the study of life" (Hoffmeyer 1997b: 940). In this perspective, semiosphere (the sphere of semiosis) coincides with biosphere (the sphere of life).[8] However, Hoffmeyer goes even further by arguing that from a semiotic point of view, "biosphere appears as a reductionist category which will have to be understood in the light of the yet more comprehensive category of the semiosphere" (Hoffmeyer 1997a: 934). "We tend to overlook the fact that all plants and animals — all organisms, come to that — live, first and foremost, in a world of *signification*. Everything an organism senses signifies something to it" (Hoffmeyer 1996: vii).

Similar conceptions of "semiobiosphere" and "biosemiosphere" have been described in semiotic literature by, respectively, Augusto Ponzio and Susan Petrilli, and Floyd Merrell. Both accounts spring from the original concept by Lotman and introduce it into the Peircean theoretical framework. Ponzio and Petrilli identify the semiosphere with biosphere and provide an integrated view of semiobiosphere that "obviously includes the *semiosphere* as constructed by human beings, by human culture, signs, symbols and

---

[8]See also Yates (1998); Kull (1998).

artifacts, etc." (Ponzio, Petrilli 2001: 264–265, original emphasis). However, they also underline that "the semiosphere is part of a *far broader* semiosphere, the semiobiosphere — a sign network human beings have never left, and to the extent that they are *living beings*, never will" (Ponzio, Petrilli 2001: 265, emphasis as in the original). Merrell (2001) follows similar assumptions.

Thus, if biosphere is defined exactly in the way as Vernadski did it — biosphere is the *matter* that is chemically changed in result of life processes — then it is not equivalent to semiosphere. However, biosphere in this sense is almost co-extensive with semiosphere, if we take the semiosphere conceptually to subsume the total sphere (or network) of sign relations (or sign processes). The set of relations that comprises everything living forms semiosphere.

Semiosphere is both a semiotic model, and an object of semiotics. This is the approach used by Juri Lotman, simply extended to cover all forms of semiosis, as biosemiotics has been doing. In this way, theory of semiosphere can be seen as a basis for general semiotics.

References

Alexandrov, Vladimir E. 2000. Biology, semiosis, and cultural difference in Lotman's semiosphere. *Comparative Literature* 52(4): 339–362.
Andrews, Edna 2003. *Conversations with Lotman: Cultural Semiotics in Language, Literature, and Cognition*. Toronto: University of Toronto Press.
Bateson, Gregory 2000 [1972]. Form, substance, and difference. In: Bateson, Gregory, *Steps to an Ecology of Mind*. Chicago: University of Chicago Press, 454–473.
Chang, Han-liang 2003. Is language a primary modeling system? On Juri Lotman's concept of semiosphere. *Sign Systems Studies* 31(1): 9–23.
Deely, John 2001. *Four Ages of Understanding: The First Postmodern Survey of Philosophy from Ancient Times to the Turn of the Twenty-first Century*. Toronto: Toronto University Press.
Deely, John 2010. *Semiotic Animal: A Postmodern Definition of "Human Being" Transcending Patriarchy and Feminism to Supersede the Ancient and Medieval 'animal rationale' with the Modern 'res cogitans'*. South Bend: St. Augustine's Press.
Hoffmeyer, Jesper 1996. *Signs of Meaning in the Universe*. Bloomington: Indiana University Press.

Hoffmeyer, Jesper 1997a. The global semiosphere. In: Rauch, Irmengard; Carr, Gerald F. (eds.), *Semiotics around the World: Synthesis in Diversity*. Proceedings of the Fifth Congress of the International Association for Semiotic Studies, Berkeley 1994. Berlin: Mouton de Gruyter, 933–936.

Hoffmeyer, Jesper 1997b. The swarming body. In: Rauch, Irmengard; Carr, Gerald F. (eds.), *Semiotics around the World: Synthesis in Diversity*. Proceedings of the Fifth Congress of the International Association for Semiotic Studies, Berkeley 1994. Berlin: Mouton de Gruyter, 937–940.

Hoffmeyer, Jesper 1998. The unfolding semiosphere. In: Vijver, Gertrudis Van de; Salthe, Stanley; Delpos, Manuela (eds.), *Evolutionary Systems: Biological and Epistemological Perspectives on Selection and Self-Organization*. Dordrecht: Kluwer, 281–293.

Hoffmeyer, Jesper; Emmeche, Claus 1991. Code-duality and the semiotics of nature. In: Anderson, Myrdene; Merrell, Floyd (eds.), *Semiotic Modeling*. New York: Mouton de Gruyter, 117–166.

Ivanov, Vjačeslav V. 1998. *Izbrannye trudy po semiotike i istorij kul'tury*. Tom I. Moskva: Shkola "Yazyki russkoj kul'tury".

Ivanov, Vjačeslav V.; Lotman, Juri M.; Pjatigorski, Aleksandr M.; Toporov, Vladimir N.; Uspenski, Boris A. 1998 [1973]. *Theses on the Semiotic Study of Cultures*. (Tartu Semiotics Library 1.) Tartu: Tartu University Press, 33–60.

Kawade, Yoshimi 1998. Imanishi Kinji's biosociology as a forerunner of the semiosphere concept. *Semiotica* 120(3/4): 273–297.

Kotov, Kaie 2002. Semiosphere: A chemistry of being. *Sign Systems Studies* 30(1): 41–56.

Kull, Kalevi 1998. On semiosis, umwelt, and semiosphere. *Semiotica* 120(3/4): 299–310.

Kull, Kalevi 2002. A sign is not alive — a text is. *Sign Systems Studies* 30(1): 327–336.

Kull, Kalevi 2005. Semiosphere and a dual ecology: Paradoxes of communication. *Sign Systems Studies* 33(1): 175–189.

Kull, Kalevi 2006. "Semiosfäär", 1982: Kommentaariks. *Acta Semiotica Estica* 3: 222–224.

Levit, George S. 2001. *Biogeochemistry, Biosphere, Noosphere: The Growth of Vladimir Ivanovich Vernadsky*. Berlin: Verlag für Wissenschaft und Bildung.

Lotman, Juri M. 1984. O semiosfere. *Trudy po znakovym sistemam* [*Sign Systems Studies*] 17: 6–23.

Lotman, Juri M. 1989. The semiosphere. *Soviet Psychology* 27(1): 40–61.

Lotman, Juri M. 1990. *Universe of the Mind. A Semiotic Theory of Culture*. Bloomington: Indiana University Press.

Lotman, Juri M. 1992. *Kul'tura i vzryv*. Moscow: Gnosis.

Lotman, Juri M. 1997. *Pis'ma. 1940–1993*. Moskva: Shkola "Yazyki russkoj kul'tury".

Lotman, Juri 2005 [1984]. On the semiosphere. (Clark, Wilma, trans.) *Sign Systems Studies* 33(1): 215–239.
Lotman, Juri M. 2009 [1992]. *Culture and Explosion*. Berlin: Mouton de Gruyter.
Lotman, Juri M. 2010. Excerpts from Universe of the Mind: A Semiotiz Theory of Culture (1990). In: Favareau, Donald (ed.), *Essential Readings in Biosemiotics: Anthology and Commentary*. Berlin: Springer, 197–214.
Lotman, Mihhail 2001. The paradoxes of semiosphere. *Sun Yat-sen Journal of Humanities* 12: 97–106.
Lotman, Mihhail 2002. Umwelt and semiosphere. *Sign Systems Studies* 30(1): 33–40.
Lovelock, James E. 1979. *Gaia: A New Look at Life on Earth*. Oxford: Oxford University Press.
Machado, Irene (ed.) 2007. *Semiótica da Cultura e Semiosfera*. São Paulo: Annablume.
Mandelker, Amy 1994. Semiotizing the sphere: Organicist theory in Lotman, Bakhtin, and Vernadsky. *Publications of the Modern Language Association* 109(3): 385–396.
Mandelker, Amy 1995. Logosphere and semiosphere: Bakhtin, Russian organicism, and the semiotics of culture. In: Mandelker, Amy (ed.), *Bakhtin in Contexts Across the Disciplines*. Evanston: Northwestern University Press, 177–190.
Markoš, Anton 2004. In the quest for novelty: Kauffman's biosphere and Lotman's semiosphere. *Sign Systems Studies* 32(1/2): 309–327.
McLuhan, Marshall 1962. *The Gutenberg Galaxy: The Making of the Typographic Man*. Toronto: University of Toronto Press.
Merrell, Floyd 2001. Lotman's semiosphere, Peirce's categories, and cultural forms of life. *Sign Systems Studies* 29(2): 385–415.
Nöth, Winfried 2006. Yuri Lotman on metaphors and culture as self-referential semiospheres. *Semiotica* 161(1/4): 249–263.
Ponzio, Augusto; Petrilli, Susan 2001. Bioethics, semiotics of life, and global communication. *Sign Systems Studies* 29(1): 263–276.
Prigogine, Ilya; Stengers, Isabelle 1984 [1979]. *Order out of Chaos*. Toronto: Bantam Books.
Sebeok, Thomas A. 1997. Global semiotics. In: Rauch, Irmengard; Carr, Gerald F. (eds.), *Semiotics around the World: Synthesis in Diversity. Proceedings of the Fifth Congress of the International Association for Semiotic Studies, Berkeley 1994*. Berlin: Mouton de Gruyter, 105–132.
Teilhard de Chardin, Pierre 1967 [1955]. *The Phenomenon of Man*. London: Fontana Books.
Torop, Peeter 2005. Semiosphere and/as the research object of semiotics of culture. *Sign Systems Studies* 33(1): 159–173.
Uexküll, Jakob von 1920. *Theoretische Biologie*. Berlin: Verlag von Gebrüder Paetel.
Vernadsky, Vladimir I. 1945. The biosphere and the noösphere. *American Scientist* 33: 1–12.

Vernadsky, Vladimir I. 1977. *Nauchnaya mysl' kak planetnoe yavlenie: Razmyshlenie naturalista*. 2-ya kniga. Moskva: Nauka.
Vernadsky, Vladimir I. 1997. *Scientific Thought as a Planetary Phenomenon*. Moscow: Nongovernmental Ecological V.I. Vernadsky Foundation.
Vernadsky, Vladimir I. 1998. *The Biosphere*. New York: A Peter A. Nevraumont Book.
Yates, F. Eugene 1998. Biosphere as semiosphere. *Semiotica* 120(3/4): 439–453.

Chapter 11
Why Do We Need Signs in Biology?

Yair Neuman

Summary. The current chapter addresses a fundamental question: Why are there sign-mediated interactions in living systems? According to Polanyi, biological hierarchies are constituted through boundary conditions. I argue that signs, or more accurately the processes of signification, function as these boundary conditions. Moreover, based on general insights from the physics of computation, I argue that the organism cannot be computed directly from the DNA without the loss of critical information. In this context, signs as boundary conditions mediate the biological construction in a way that prevents the loss of information and destabilization of the DNA.

Introduction

In the minds of scholars and lay persons alike, the concept of the sign is usually associated with the linguistic realm, in which signs are used as a vehicle of communication between human agents. However, in the most general sense, a sign can be considered to be a "carrier of meaning" and as such it exists in biology. The reason I put the expression "carrier of meaning" in quotation marks is that meaning cannot be carried as if it were an object. This is a misleading metaphor. Meaning always involves a response and is thus an activity rather than an object. This idea is clearly presented by Holquist: "lack of water means nothing without the response of thirst. [...] It is still the case that nothing means anything until it achieves a response" (Holquist 1990: 48). For example, the meaning of a monstrous face staring at us from the dark is the response of flight. No meaning is encapsulated in the face. The meaning of a molecule's being identified as an antigen can be comprehended only through the immune response (Cohen 2000). No meaning is encapsulated in the molecule. Why do I emphasize the idea of meaning as a response? For two reasons: first, to dismiss the misconception that meaning is encapsulated in the message; and second, to introduce the idea that the sign is not a "carrier of meaning" but a trigger for meaning-making (Neuman 2004).

A variety of "meaning triggers", such as mRNA, cytokines and hormones, are evident in organisms, specifically those that are considered to be "higher-order" in terms of various complexity measures. Those triggers may be studied through biological, physical or chemical lenses. However, as signs they may also be approached from a more general semiotic perspective that deals with issues of signs and signification. In this context, a semiotic analysis may at least provide theoretical biology with some interesting suggestions regarding signs and signification in living systems.

The relevance of a semiotic analysis to biology was originally developed in the first half of the 20th century by Jakob von Uexküll (1982 [1940]). His work made it clear not only that the concepts of "sign" and "signification" are relevant to theoretical biology but that they are indispensable for understanding the unique nature of living systems as opposed to matter. The aim of this chapter is not to propagate or to review this theoretical perspective, a task that has been done in other places (Hoffmeyer 1996; Kull 1999; Neuman 2008; Sebeok 2001), but to address a fundamental question in theoretical biology from a biosemiotic perspective: *Why are sign-mediated activities evident in organisms?* This question will be addressed specifically with regard to the genetic realm of higher-order organisms.

Why do we need signs in biology?

Any inquiry into the foundations of semiotics should address the question of when and why a direct encounter between biological components/systems is possible and when and why sign-mediated interaction is a must. This question is far from mere philosophical casuistry; its implications are evident in biology. Certain interactions, specifically at the atomic and molecular scale of analysis, do not seem to be sign-mediated. For example, the interaction of enzyme and substrate may be described according to the lock-and-key metaphor (Clardy 1999) as a direct structural encounter between two entities through non-covalent forces. No signs are evident in this encounter. In other cases, such as the transformation from DNA to proteins, mediation through mRNA is clearly evident.

One may consider the question of why the DNA cannot be used directly to synthesize proteins without the mediation of RNA. Answers to this question — or, more accurately, scholarly speculations — can be provided indirectly from functional or evolutionary perspectives. However, those answers do not directly address the question and the reason for the existence of semiotic mediation in the realm of living organisms. For example, Francis Crick argued that life (or at least genetic replication) started with RNA. DNA, which is a more stable molecule and is better for long-term storage of genetic information, came later. This suggestion concerns the evolutionary primacy of the RNA but does not explain why RNA is a necessary mediator for protein synthesis (for a recent explanation see Neuman, Nave 2008). Another explanation is that because there is usually only one copy of any particular gene in the cell, the movement from DNA to protein is much more rapidly mediated through the RNA (Alberts *et al.* 1998). In other words, the RNA allows "synthesizing the required amount of protein much more rapidly than if the DNA itself were acting as a direct template for protein synthesis" (Alberts *et al.* 1998: 212).

Another possible explanation for the existence of RNA is a functional one. The DNA cannot leave the nucleus. However, information must be carried out from the nucleus to the ribosome. In this sense mRNA clearly functions as a sign since it functions as a "carrier of information" from one system to the other. These answers are either evolutionary or functional but do not address the question that opened this section from the most basic biosemiotic perspective. In order to answer this question from a basic biosemiotic perspective, I would like to introduce the term "transmutation".

Life's irreducible structure

In his seminal paper on translation, the linguist Roman Jakobson (1971) made an important distinction between two concepts: *translation* and *transmutation*. While translation concerns a transformation from one semiotic system to another, transmutation involves transformation from a verbal to a non-verbal semiotic system. The term "transmutation" may be enlarged to include *transformation between semiotic and non-semiotic systems*, and here I will use it in this specific sense. That is, the term transmutation will be used to denote the transformation between a systems of signs to another system which is not a systems of signs.

Why is the concept of transmutation relevant to our inquiry? In order to explain this relevance, we should move on to Polanyi's classic paper "Life's irreducible structure" (1968).

As a scientist, Polanyi did not doubt that organisms are composed of matter that may be described in physical terms. However, the main point in his paper is that what is important for our understanding of living systems is not matter as such but the "structure of boundary conditions" or the "restrictions" that constitute the "biological hierarchies" of which organisms are composed. To use an analogy, "when a sculptor shapes a stone and a painter composes a painting, our interest lies in the boundaries imposed on a material, and not in the material itself" (Polanyi 1968: 1308).

The most important implication of Polanyi's argument for the current chapter is as follows: the biological hierarchy is composed of various levels that can be described in physical terms. If these levels interact through boundary conditions to constitute the living organism, then the transformation from one level to another of the biological hierarchy is far from trivial. This transformation cannot be simply reduced to the laws governing the levels themselves. It is a transformation from one form of organized matter to another form of organized matter that should be mediated according to laws that cannot be reduced to the laws of physics.

My first suggestion is that signs are the vehicles for transmutation, for transforming an input — one form of organized matter, into another form of organized matter.

In other words, in biology the sign is primarily a vehicle for moving between subsystems and levels and it serves as a boundary condition (Polanyi 1968) for the transformation between different layers of matter (i.e., non-semiotic systems).

Moreover, the biological hierarchies discussed by Polanyi may be considered the output of a computation process performed on the input, for instance the genome. In this argument, which is the bread and butter of modern biology, a computation process should be understood in its most general sense as a procedure that uses an algorithm to produce an output from a given input. However, according to insights gained in the physics of computation (Bennett, Landauer 1985), any process of computation involves a loss of information. In other words, *if higher levels of the organism's hierarchy are the computational output of lower-level DNA, then the result is inevitably a loss of information in the shift between the levels.*

My second suggestion is that signs function not only as boundary conditions but as a unique interface that mediates the computation of the genomic input without causing a loss of information.

To sum up, my main argument is that a major function of signs, at least in higher-order organisms, is to mediate between different layers/systems of biological hierarchies and to compensate for the irreversible process of computation evident in the biological construction.

Transcending the view "from within"

Let me begin my discussion by asking a very general question: What is the minimum condition for semiotics, i.e., for the use of signs in a system? The answer is clear: a distinction. That is, at the heart of any semiotic activity we must assume the existence of differentiated states that are being subject to signification. In a world of total chaos no names are given. In the book of Genesis, for instance, differentiated states (fowls, beasts, etc.) were created first, and only afterwards did Adam name them.

If we accept the idea that discrete physical states are the primitive units of matter, then those differentiated states should be considered the minimum units of a semiotic analysis. However, the existence of differentiated physical states at the base of our ontological hierarchy does not necessarily imply that there is a contemplating mind interpreting these differentiated states as signs. After all, differentiated physical states existed long before organisms started populating the earth. Why do I emphasize this point? Because a physical distinction can not be truly used as a unit of signification. Any physical difference/distinction is a singular event without the existence of a contemplating device that may convert it into a more general and abstract instance of a class that may be communicated across domains (i.e., a sign). Deleuze (1990) even coined the term "repetition" to describe this ontological category of a "difference without a concept", or, to borrow from phenomenological jargon, a difference "in and for itself".

Physical states are pure singularities, a property derived from the fact that they occupy different positions in space–time. For example, as a physical entity, each cat is a unique creature with its particular position in space and time. It is what Peirce described as the "dynamical object". Only the ability to *group* the various instances under the concept "cat" makes it possible to approach the singular cat from a general and abstract perspective and to communicate this general perspective across space and time as a sign. Following this line of reasoning, I would like to define a sign as follows:

A sign is the name we, as outside observers, give a concept — a functional generality — that is communicated across realms.

Let me illustrate this idea. Each of the four DNA nucleotides is a singularity, since it occupies a unique position in the linear sequence of the DNA. However, when transcribed into mRNA these nucleotides lose their singularity to become a part of the DNA triplets known as codons, each signifying/specifying the synthesis of a specific type of amino acid. A codon is a sign — a communicated functional generality. For example, in the standard code, the codon CUC specifies leucine, CGU specifies arginine, and CAU specifies histidine. Codons are signs in another sense as well: that they may signify different things to different observers. Although in the standard code CUU specifies leucine, for yeast it specifies threonine! In other words, there is no one-to-one correspondence between the sign and response it invites (i.e., its meaning). This is a general characteristic of signs, and biological signs are no exception. Codons are also signs in another important sense. A sign may turn out to be a signified realm (i.e., a source of signification) itself. Codons may also turn out to be signified realms themselves. To reiterate, codons are used for the synthesis of amino acids through the mediation of transfer RNA (tRNA). Through the anticodon region, the tRNA adheres to the codon, and through a short single-strand region at the $3'$ end of the molecule, the amino acid that matches the codon is attached to the tRNA. In this case, it is the tRNA that turns out to be a signifying process.

Now we start to understand why signs as communicated functional generalities are necessary in biological systems. Physical systems as conglomerates of singularities are embodied in local interactions within the system, while biological systems are capable of synthesizing proteins because they can "transcend" the singularity through the generality and the communicability of signs/codes. Indeed, transcending the singularity of matter is a constitutive dynamic of living systems that is possible only through signs. Let us move on to our next station as we continue our journey.

Information as a "difference that makes a difference"

As Bateson (2000) argued in his seminal essay "Form, Substance and Difference" (originally published in 1970), the realm of the living is a realm in which effects (i.e., responses) are brought about by information that he uniquely defines as a "difference that makes a difference". That is, in the realm of the living, pure differences — discrete physical states — are not enough. Only differentiated states (i.e., differences) that are actively differentiated on a higher level of analysis (i.e., a difference that makes a difference) can constitute the realm of the living.[1]

The importance of Bateson's unique definition of information for biology cannot be underestimated. For instance, one should notice that the existence of an "observer", is built into Bateson's definition of information. A difference may "make a difference" only to something or someone (e.g., molecular machinery). This is the context in which I elsewhere suggested that living systems are "meaning-making" systems rather than "information-processing" devices (Neuman 2008).

Why do we need Bateson's and Polanyi's ideas to understand sign-mediated activity? The answer I would like to give is somehow surprising, although it is rooted in an idea that appears in slightly different form in Polanyi's paper. We may suggest that the realm of the living is a not simply one of complex combinations of particles of matter. It is a realm in which matter is *actively* transformed

---

[1] For a recent elaboration of this idea see Neuman, Tamir (2009).

by molecular devices into informational content (i.e., a boundary condition or "a difference that makes a difference"), which is recursively used for the construction and constitution of the organism. In other words:

An organism is constituted as a recursive hierarchy that turns differences (i.e., discrete physical states) into informational content for the active production of other discrete physical states (i.e., its self-creation, or autopoiesis).

That is, novelty of biological construction through informational content is the living system's *sine qua non*. However, the physics of computation teaches us a lesson about the prices of this process, prices/constraints that should be taken into consideration through sign-mediated activity. This lesson will add another layer to our understanding of signs in living systems.

Machines of oblivion

Novelty, as is evident in the realm of the living, results not only from turning physical differences into informational content, but also from *erasing information* in order to create qualitatively new structures. That is, at the heart of emerging biological structures is the idea that something is necessarily gained and something is necessarily lost when we shift between levels of the biological hierarchy (Neuman 2007). This state of affairs is crystal clear for the embryologist who observes the apoptosis of cells as a natural process of embryonic development.

The famous novelist Jorge Luis Borges (2000: 183) beautifully epitomizes the close relation between novelty and oblivion in one of his stories:

> Solomon saith: *There is no new thing upon the earth.* So that as Plato had an imagination, *that all knowledge was but remembrance*; so Solomon giveth his sentence, that all novelty is but oblivion.

As Borges, who attributes the above piece to Francis Bacon, suggests, "all novelty is but oblivion". This idea might seem too poetic for biologists but it may become comprehensible when examined in terms of the physics of computation. Indeed, a recent paper on the physics of computation was entitled "The physics of forgetting" (Plenio, Vitelli 2001).

As argued by Bennett and Landauer (1985), a process of computation involves the loss of information; they define computation as a process in which an "output" is produced from an "input" through an algorithm, and "information" is considered in the most general sense of differentiated states.

Let me explain the loss of information with a simple example: The arithmetic expression $1 + 1 = 2$ involves a process of computation in which the binary operation of adding the inputs "1" and "1" produces the output "2". Why and in what sense does this process involve the loss of information? There are physical and computational ways of describing and explaining this loss. One way to think of information erasure is in terms of computational irreversibility. A logical gate such as OR *is irreversible if, given the output of the gate, the input is not uniquely determined.* For example, the logical gate NAND is intrinsically irreversible. If the output of the gate is 1 then the input could have been 00, 01, or 10 (Nielsen, Chuang 2000: 153). The same process is evident with regard to the simple arithmetic expression previously presented. Without getting into the particularities, qualifications and difficulties of this argument, we should acknowledge the consensus that computation necessarily involves the loss of information in the most general sense of "differentiated states" (Bennett, Landauer 1985). In other words, when a difference is turned into a difference that makes a difference, some information is lost. The same argument holds in biology. Each cell in our body contains the same DNA. However, the computation of the biological hierarchy from this information source clearly involves a loss of differentiation at the different stationary states of the construction process, as is evident in the transformation from DNA to the RNA.

Reductionism is impossible because we are irreversible

The idea presented above has crucial implications for understanding reductionism in biology from a novel perspective. Let me explain this argument: At this point in our discussion it is clear that the fact that higher-order organisms do not easily lend themselves to reducible analysis is not due to the failure of brilliant minds to accomplish this task. It is an intrinsic property of multi-level biological systems that are clearly irreducible due to the irreversibility of their construction process. In other words, *biological systems involve an irreversible process of computation* and therefore simply tracing the biological output back to its constitutive elements is impossible. Information is necessarily lost in the process of constructing the organism and this loss is the *sine qua non* of the construction process. The question is: So what? Even if the construction of the organism involves a certain loss of information, as implied by the physics of computation, what has the sign got to do with it?

The digital and the analogical

In my presentation above, *I was deliberately misleading*. Living systems are not typical computational machines. They are not Turing machines but machines of interaction (see Neuman 2008). They are irreversible in the sense that genomic information is necessarily lost in the computation of the biological hierarchy. However — and this is the important point — they are computational machines that jealously preserve their input (the DNA) in every cell of their bodies. The most important implication of this observation is that a direct computation from the DNA would have resulted in a loss of information and the diminution of the organism's most important "text" of stability. In other words, a direct, unmediated computation from the DNA would have been incompatible with the essence of the DNA as a source of informational stability. The solution is a sign-mediated computation. The DNA, a relatively stable source of information, remains intact as an input; it is only copied according to a one-to-one correspondence that preserves its identity and its informational content. On the other hand, the genome is continuously interpreted through sign-mediated activity that leads to the constitution of the organism.

Readers familiar with the biosemiotics literature may immediately associate this thesis with the "code duality" of Hoffmeyer and Emmeche (1991). However, long before code duality was introduced, Gregory Bateson made the clear-cut distinction between *digital* and *analogue* "modes of communication". After carefully reading Bateson, I believe that the most important feature of the digital code is that it operates *on the same level of logical analysis* (Bateson 2000: 291). In Bateson's terms (2000: 140, 291), DNA serves as a digital code of communication. This is why the only direct operation in which DNA is involved is *copying*, an operation that takes place as a one-to-one mapping on the same logical level of analysis. Copying preserves the DNA's digital code: guanine is guanine is guanine! Copying, however, can create nothing new. To construct an organism we must transcend the "view from within", we must transcend the digital code, and therefore the analogue code must be used. However, any form of coding directly from the DNA would have resulted in the erasure of the informational input. Remember the lesson we learned from the physics of computation? You can't beat city hall! Nature's solution is the sign. The sign is a functional generality that does not threat the stability of the DNA. The information represented by the DNA is not erased. It is "interpreted" in the most basic sense of the term. The use of the term "interpretation" in the biological sense is not an intellectual whim. Biological interpretation, like linguistic interpretation, is a sign-mediated process. I use the term "interpretation" to denote:

The production of different functional generalities from one set of distinctive features.

The RNA codons are the result of an interpretation process. It uses the text as a source of reference (i.e., as an input) but does not get rid of it. It is a unique form of computation in which the stability of the DNA is assured and at the same time transcendent in a way that allows for the construction of novel biological forms. This is the main argument of this chapter.

Conclusions

Organisms are machines of novelty and oblivion. Through boundary conditions they transcend the "view from within" and constitute the living organism as sign-mediated matter. However, living systems have to strike a delicate balance between novelty and oblivion.

Organisms retain their genes as much as possible. So strong is this tendency to preserve the "holy scriptures" of the genome that a whole theory, that of the "selfish gene", was constructed around this observation (Dawkins 1976). However, organisms do not accept the fundamentalist idea of rejecting the present in favor of ancient holy texts either. Like Talmudic sages, organisms interpret the ancient text of the genes for their autopoiesis and survival in the present.

Reading the distant and ancient text of the genes creates a problem for the organism, which is solved by the "dual code". This problem is encountered by philologists, too. The Spanish philologist José Ortega y Gasset (1959, quoted in Becker 2000) began one of his seminars by discussing the "difficulty of reading" (Becker 2000). "To read a distant text", he wrote, "distant in space, time, or conceptual world — is a utopian task". This task is no different from that facing an organism that is reading the distant text of its genes. The task, adds Ortega y Gasset, is one whose "initial intention cannot be fulfilled in the development of its activity and which has to be satisfied with approximations essentially contradictory to the purpose which had started it". Commenting on this statement, the linguistic anthropologist Anton Becker says, "In that sense the activity of language is in many particular ways utopian: One can never convey what one wants to convey" (Becker 2000: 298). As Ortega y Gasset puts it, it is *deficient* in the sense that it says less than it wishes to say, and it is *exuberant* in the sense that "it says more than it plans". This "utopian" characteristic of language is a source of flexibility that results from signs that are simultaneously deficient and exuberant. A sign always "says" less than it "plans" in the sense that as a functional generality it may serve different functions in different contexts. It is exuberant and always "says" something it did not "plan", in the sense of a generality that transcends the level of logical analysis from which it emerges. Codons are simple instances of these properties, but this certainly does not exhaust the list of sign activities in biological systems.

References

Alberts, Bruce; Bray, Dennis; Johnson, Alexander; Lewis, Julian; Raff, Martin; Roberts, Keith; Walter, Peter 1998. *Essential Cell Biology: An Introduction to the Molecular Biology of the Cell*. New York: Garland Publishing Inc.

Bateson, Gregory 2000. *Steps to an Ecology of Mind*. Chicago: The University of Chicago Press.

Becker, Anton L. 2000. *Beyond Translation: Essays Toward a Modern Philology*. Ann Arbor: The University of Michigan Press.

Bennett, Charles H.; Landauer, Rolf 1985. The fundamental physical limits of computation. *Scientific American* 253(1): 48–56.

Borges, Jorge Luis 2000. *Collected Fictions*. (Hurley, Andrew, trans.) London: Penguin.

Cohen, Irun R. 2000. *Tending Adam's Garden*. New York: Academic Press.

Clardy, Jon 1999. Borrowing to make ends meet. *Proceedings of the National Academy of Sciences of the United States of America* 96(5): 1826–1827.

Deleuze, Gilles 1990. *The Logic of Sense*. (Lester, Mark, trans.) New York: Columbia University Press.

Dawkins, Richard 1976. *The Selfish Gene*. New York: Oxford University Press.

Hoffmeyer, Jesper 1996. *Signs of Meaning in the Universe*. Bloomington: Indiana University Press.

Hoffmeyer, Jesper; Emmeche, Claus 1991. Code-duality and the semiotics of nature. In: Anderson, Myrdene; Merrell, Floyd (eds.), *Semiotic Modeling*. New York: Mouton de Gruyter, 117–166.

Holquist, Michael 1990. *Dialogism: Bakhtin and His World*. London: Routledge.

Jakobson, Roman 1971. On linguistic aspects of translation. In: Jakobson, Roman, *Selected Writings II. Word and Language*. The Hague: Mouton, 260–266.

Kull, Kalevi 1999. Biosemiotics in the twentieth century: A view from biology. *Semiotica* 127(1/4): 385–414.

Neuman, Yair 2004. Meaning-making in the immune system. *Perspectives in Biology and Medicine* 47: 317–328.

Neuman, Yair 2007. Immune memory, immune oblivion: A lesson from Funes the memorious. *Progress in Biophysics and Molecular Biology* 92: 258–267.

Neuman, Yair 2008. *Reviving the Living: Meaning Making in Living Systems.* New York: Elsevier.

Neuman, Yair; Nave, Ophir 2008. On the semio-mathematical nature of codes. *Biosemiotics* 1(1): 99–113.

Neuman, Yair; Tamir, Boaz 2009. On meaning, self-consciousness and quantum physics. *Journal of Cosmology* 3: 540–547.

Nielsen, Michael A.; Chuang, Isaac L. 2000. *Quantum Computation and Quantum Information.* Cambridge: Cambridge University Press.

Plenio, Martin B.; Vitelli, Vincenzo 2001. The physics of forgetting: Landauer's erasure principle and information theory. *Contemporary Physics* 42(1): 25–60.

Polanyi, Michael 1968. Life's irreducible structure. *Science* 160: 1308–1312.

Sebeok, Thomas A. 2001. Biosemiotics: Its roots, proliferation, and prospects. *Semiotica* 134(1/4): 61–78.

Uexküll, Jakob von 1982 [1940]. The theory of meaning. *Semiotica* 42: 25–82.

Part III

Conversations

Chapter 12
Between Physics and Semiotics

Howard H. Pattee and Kalevi Kull

Summary. In this dialogue, we discuss the contrast between inexorable physical laws and the semiotic freedom of life. We agree that material and symbolic structures require complementary descriptions, as do the many hierarchical levels of their organizations. We try to clarify our concepts of laws, constraints, rules, symbols, memory, interpreters, and semiotic control. We briefly describe our different personal backgrounds that led us to a biosemiotic approach, and we speculate on the future directions of biosemiotics.

We started this conversation standing at the base of Massachusetts' highest mountain. The forest on the top was hidden from our sight by clouds. We talked on several biosemiotic themes, which we develop further here; but where they lead us is unpredictable. That is life.

The regions for life in the physical world

K.K. The first problem we need to solve is evidently to demonstrate how the possibility of *choosing one's path* — a characteristic feature of all life — *can be embedded into the picture of a physical world* which is based on inexorable physical laws. Everything in the world (at least that can be measured) is consistent with physical laws. Yet there is what you have characterized as *open-ended evolution*.[1] Would you agree if we call it — equivalently — a *freedom to establish new rules*?

---

[1] For example, Pattee (1988: 69).

H.P. Yes, it would certainly include establishing new rules, but by *open-ended* I want to include *any* emergent structure, function or behavior that can be imagined — or perhaps even behavior that we can't imagine because of the limitations of our current brains. We can't predict what novelties evolution might produce.

K.K. The open-ended evolution includes then two distinct properties. (1) An immense[2] number of potential forms, and (2) a basic unpredictability of the paths evolution will take. These features, accordingly, apply to biological evolution and do not apply to evolution in the non-living world.

H.P. The physical basis of the immense number of forms is a consequence of the immense number of linear sequences of material units that laws cannot distinguish because of their similar energy or similar stability. This is the genetic memory. Only some form of "frozen accident" or higher level selection process affects which memory sequences survive over time. Not only are the initial sequences unpredictable, but their physical structure appears to be largely arbitrary. Natural selection is also unpredictable because of its complexity and the indefinite time period over which selection continues to work.

The most obvious, and I would say the most important, similarities of genetic language and human natural, formal, and computer languages is their expression by such discrete, linear strings using only a small, materially arbitrary alphabet. It is just these properties that allow simple and reliable writing, reading, and storage in a memory that is lawfully undetermined, and that allows practically unlimited information capacity.

K.K. The consistency with physical laws means that everything is *dependent* on the laws — none of the biological or mental processes is inconsistent with any physical law. However, as you say, this does

---

[2] With this term I would refer to Walter Elsasser (1998: 49ff) who emphasized the role of immenseness in this sense as a characteristic feature of life.

not mean that everything is *determined* by the laws.[3] The "regions of indeterminacy" are supposedly those in which life can establish itself and evolve. Is it possible to describe these regions of indeterminacy and how they arise?

H.P. The inexorable character of physical law is often misunderstood to imply determinism. This is not the case. There are innumerable structures in the universe that physical laws do not determine. It is also important to understand why *lawfully indeterminate* does not mean *physically indistinguishable*.

Since all the basic laws of physics are expressed in terms of energy, systems with two or more states with the same energy are lawfully indeterminate. However, in many cases we can distinguish these states by measurements of their initial conditions. These law-equivalent states are often called degeneracies or symmetries.

A common example is chirality, or left and right handedness. Chemically, amino acids and proteins can be left or right handed, and they cannot be distinguished by the laws that they both obey. Nevertheless, most types of biochemicals in living organisms must stick with one or the other.

This is like our driving on one side of the road. Either side would work just as well as the other, but we have to choose one for traffic to function efficiently. Such symmetry-breaking events that persist for structural, functional, or selective reasons are appropriately called "frozen accidents".

The most important energy-degenerate structures for life and language are the linear strings of discrete units like nucleic acids and the strings of symbols like the words on this page that form a memory. It is just because of the immense numbers of these energy-indeterminate but structurally distinguishable strings of symbols that their information storage capacity is open-ended. These distinguishable sequences and the information they contain

---

[3]Pattee (2008: 151).

are not determined or restricted by natural laws; but their relative permanence is the result of frozen accidents, and natural or cultural selection in specific environments. Most linguistic conventions are probably frozen accidents. It is possible that the genetic code and consequently life itself began as a broken symmetry that became a frozen accident.

K.K. Isn't "redundancy" a more precise term than "degeneracy" in these cases?

H.P. The word "degeneracy" is physical jargon that is not equivalent to "redundancy". More precisely, linear copolymers are "near-degeneracies", meaning that their stabilities or lifetimes are nearly the same as long as they remain linear and isolated. So far this is just a "meaningless" physical necessity that allows an unlimited variety of sequences.

Degeneracy is more closely related to what physicists call a symmetry where any change of sequence order does not change the law-based description. Degeneracy has nothing to do yet with semiotics or potential functions where "redundancy" may have meaning.

K.K. Still, it seems to me, we have not yet entirely explained how the living systems — or sign processes — escape from the determinism of the physical laws. Because the existence of energetically degenerate states yet does not mean that what will happen will not be determined by the initial conditions — as for instance in case of a growth of a nucleotide strain, the choice of the next nucleotide is not determined by the previous nucleotide; however, it can be determined by a movement of the nucleotides around (e.g., the one that will reach the endpoint of the strain first would stay there).

In other words, in order to explain the appearance of semiotic freedom[4] the existence of law-equivalent events is necessary, but

---

[4] The concept of semiotic freedom is central to Jesper Hoffmeyer's writings (for example, Hoffmeyer 2008).

this cannot be achieved on a solely molecular level. It is necessary to demonstrate the emergence of non-determined regulation by boundary conditions. Otherwise the freedom is basically illusionary, as, for instance, Daniel Dennett would claim.

H.P. The concept of absolute determinism as envisioned by Laplace and philosophers like Dennett, has turned out in physics to be an unsupportable and unproductive way of thinking. Determinism is an untestable metaphysical concept. First of all, measurement processes are irreversible and therefore dissipative and subject to error, so determinism is not empirically verifiable. All the fundamental laws are consistent only with a probabilistic universe. We have enough "freedom" just because of the undeterminable or equivalent probabilities of many structures, like polymer sequences.

There is also plenty of freedom just at this molecular level to allow brains to make choices because all brain function is dependent on the molecular level. As Arthur Eddington (1929: 260) noted long ago: "There is nothing to prevent the assemblage of atoms constituting a brain from being of itself a thinking object [with 'free will'] in virtue of that nature which physics leaves undetermined and undeterminable".

K.K. You have used the term *constraint* as a central notion in your writings. How should constraint be defined?

H.P. In physics a constraint is a local structure that limits the motions of otherwise "free" particles that are governed only by the laws of motion. However, the concept of constraint is also used to describe levels of hierarchical organizations. Generally speaking, each higher level requires a constraint that is described by fewer observables than the lower level description. More precisely, a constraint is an *alternative simplified description* of structures that are not usefully described by the behavior at a more detailed lower level.

A simple example is a closed box that limits the detailed motions of the gas molecules inside. The box itself is also made of molecules, but they are constrained by chemical bonds to form a solid structure. So we simplify the description of the box by describing only its geometric boundaries, and we ignore the detailed molecular structure of the box itself. Such constraints are also called "boundary conditions".

A more complicated example is an internal combustion engine. The entire engine is made of molecules, but they are so rigidly constrained as solid parts that we can usefully describe the engine's motion by just one rotational degree of freedom. Engines constrain the gas molecules in the cylinders so that, by coupling to several higher levels of fixed and moving constraints, it does useful work.

A very complicated example is the computer that at the lowest level simply constrains the flow of electrons. At a higher design level these constraints are described as circuits, memories and gates. However, when we use a computer we ignore this hardware level of constraint because it is more practical to control its behavior at a higher level of *symbolic constraint* we call a code or program. A reasonably complete understanding of a modern computer requires different descriptions and languages for at least six levels of constraints.

Biosemiotics covers even more levels of constraints, from the chemical bonds that constrain gene sequences, enzyme dynamics, and cell membranes, to the matter–symbol transition of the structural genes and the epigenetic controls of development, and finally to the nervous system architecture and the brain. Consequently our biosemiotic models require many different levels of descriptions. Failure to recognize that these different levels of descriptions are necessary and *complementary* often causes useless arguments over which is the "best" description.

By contrast with the computer, the organism itself must develop almost all of its higher level constraint structures under the supervision of the genetic description. The genetic constraints harness the *self-organizing* physical laws with great efficiency, like protein folding and the self-assembly of components, all of which follow energy-dependent laws. In other words, a relatively few genetic constraints control a large number of energy-based physical actions and constructions. As we are now learning from gene sequencing, the simple structural genes are only a small fraction of the genome. Most of the genome is made up of control sequences that are coordinated by extremely complex linkages. How this coordination arises is the key problem of evolution and development (Pattee 1971a).

K.K. Thus, there are constraints both in the non-living and the living world. But aren't these constraints radically different? As we can observe, the constraints in the living world are, (a) fundamentally individual, due to the individuality of each organism, and (b) mutual or reciprocal, due to the communication processes that occur between any living beings, between the cells, and between the organisms. As a result of the individual and mutual constraints, the relations become established between the living systems — the relations (I would see these also as rules, or codes) that might be unpredictable from the physical laws.

H.P. Yes, living and non-living constraints are radically different for the reasons you give. Non-living constraints are not constructed from heritable memory that persists by natural selection. Living constraints occur in individuals with a memory. Genes are the memory that define the individual. As Hippocrates recognized, your conscious individual self is memory in your brain. All your other organs can be transplanted without changing your individual awareness. The same is true of the genes at the cellular level.

K.K. A living system can establish constraints and do work in this way. Via doing work, it can then build whichever structures, both useful and just for fun, or also in a "let's see what comes out of it" way. The work done with the help of constraints is using physical processes, which means that no freedom from the laws, no indeterminacy is really required for this part of the life process. Freedom from laws (in the sense of law-equivalence, or indeterminacy), however, as it seems, is necessary in order to make choices, i.e. to behave in a way that would not be predictable by any computational means.

H.P. To be more precise, you never have "freedom from laws" but only freedom from initial or boundary conditions. You have to make a clear distinction between laws and constraints. Laws are universal and inexorable. Nothing is free of laws. Constraints are local structures that obey laws but are not determined or predictable by laws. Memory is a special type of constraint that can alter or control the lawful course of local events. Polanyi's (1968) phrase "harnessing the laws" is apt.

It is only memory constraints that allow an organism's heritability, variation and natural selection. At the cognitive level, it is only by consulting our memory that we feel we are making choices. A sudden response to a stimulus, like a loud noise, does not feel like a choice.

K.K. Nevertheless, we may probably think of common free behavior also without any inclusion of law-equivalent states. For instance, if a behavioral act is a habit-based search for an object represented by its memory, driven by an organism's need and taking into account the umwelt around it — it is not obvious that any law-equivalent state is required for such a behavior.

H.P. Exactly. It is just because it is law-equivalent that law-based thinking is irrelevant. Semiotic expression is free of physical laws. The existence of any memory requires many law-equivalent states. In fact, the information capacity of a memory is defined in terms of the number of law-equivalent (equiprobable) states.

K.K. A behavioral act may result in some learning, which means a slight change of memory, and thus the behavior will be fully individual, and also unpredictable, because the response cannot be calculated — exactly analogically to the complete function of an enzyme that also cannot be calculated. Thus, a question still is whether a complex individuality is not already sufficient to provide all behaviors the organisms may have.

H.P. Again I would say that any "complex individuality" is defined only by its memory, and therefore such memory-controlled individuals would be capable of evolving or learning many forms of behavior.

K.K. A particular constraint can be produced deterministically, like the shores of a river as the river is shaping them. But if there is a system of constraints, in which the constraints mutually rebuild each other, whereas the reproduction of the constraints is based on a non-exact mutual recognition, then an identity can arise, which turns out to be quite independent from microprocesses.

For instance, in a population of biparentally reproducing organisms each individual is genetically different from any other, but they recognize each other when producing offspring and thus form a species that holds itself.

This is like an ongoing communication, in which the communicants reciprocally constrain each other and thus the self-identity of the communication process is kept. Life is probably just this kind of general communication process.

In order to get life running, what is required is an inheritance mechanism, i.e., memory — the one that consumes energy in order to rebuild itself; the inheritance mechanism[5] obviously has to include semiosis, because it has to find and recognize its building blocks. And the inheritance mechanism is nothing more than a general self-supporting communication mechanism, as I just tried to describe it.

H.P. I would agree that even the simplest reproduction requires the communication of information from parent to offspring. All multicellular development is also dependent on communication

---

[5]Inheritance is meant here in a broad sense, like, for instance, Jablonka and Lamb (2005), who include in it the epigenetic, genetic, behavioral, and symbolic inheritance.

between cells. I would still argue that some form of memory, not necessarily discrete symbol strings, is the source of all heritable information. Where would a semiotician say symbols are located when not in use?

K.K. That's right. Where memory, of course, is not just a structure but a correspondence (i.e., a *relation* between structures) that is modified and conveyed in semiosis.

Thus biosemiotics is the field that not only tackles the mind–matter problem, but also addresses the problem of complementarity of semiotic and physical descriptions at all levels. There is a whole series of problems of the "symbol–matter" type that you have listed in your writings. Can you describe these?

H.P. The amazing property of symbols is their ability to control the lawful behavior of matter, while the laws, on the other hand, do not exert control over the symbols or their coded references. It is just for this reason that evolution can construct endless varieties of species and the brain can learn and create endless varieties of models of the world.

That is why organisms and symbol systems in some sense locally appear to escape the global behavior of physical laws, yet without ever disobeying them. Fully understanding this power of symbols over matter at *all* evolutionary levels is what I call "the symbol–matter problem".

The four most notorious symbol–matter levels are the genetic code in biology, pattern recognition and sensorimotor control in nervous systems, the measurement and control problem in physics, and the mind–body problem in philosophy.

K.K. It occurs to me that any true model of semiosis has to include in itself the "symbol–matter problem" as you call it. The models of sign that don't include it may be useful in certain cases, but in order to be a model of semiosis, i.e., of sign process, the inclusion of the symbol–matter problem is inescapable.

H.P. I think the "symbol–matter problem" is maybe not the best name because it is a triadic relation. The symbol and matter must be connected by an intepreter (Peirce's "system of interpretance"). Following the physicists' use of "cut" to separate the measurement from what is measured, I have also called the necessary separation of symbol and referent the "epistemic cut", which is also a triadic relation that must comprise the interpreter.

Both these phrases appear to evade the problem because symbol, matter, and cut are relatively simple to describe compared to what is necessary to describe an actual measurement process or any system of interpretance. I have said, along with most biosemioticians, that *the simplest system of interpretance is the living cell* (Pattee 1969). I have also suggested that the enzyme constitutes the simplest functional measuring device (Pattee 1971b). Only if the enzyme recognizes (measures) its substrate by binding does it function as a specific catalyst. Furthermore, the relation between its substrate recognition and its catalytic function is not determined by laws but only by virtue of its genetic construction.

K.K. I think we need a special term to mark the connections or structures that are made specifically by semiosis, i.e., via a semiotic control. These are the pieces of semiosis "left behind", the fractions that are produced as parts of relations or codes, or of memory. In the cultural sphere, these are usually called "artefacts", but as far as I know there is no general term for this in biology or in physics.[6] These are the structures made when using the physical indeterminacy — like the proteins that are built by ribozymes, or the nests built by birds. Most living matter (as chemical structures) is such, and also

---

[6] Except a proposal made by M. Barbieri, who proposes simply to extend the term "artefact" over everything made (or manufactured) by life. Another, but different, approach is developed by J. Deely, who is extending the term "object" to anything that is either recognized or produced; since, however, the objects are (as Deely argues) always a part of the action of signs, this leads him as a result to extend the semiosis to occur in the non-living world.

what remains after an organism has died, so most of the material in ecosystems belongs to this, because it is made via semiotic control.

What is interesting with these "products of semiosis" — while these are often very different from the structures that appear without any semiotic control in the non-living world, these may also be indistinguishable from the latter. An oligopeptide produced by a cell may be indistinguishable from an oligopeptide which has formed by a stochastic condensation of aminoacids; likewise a replacement of some stones on a stony seashore may be indistinguishable from a replacement caused by waves, or even CO_2 synthesized by cells via respiration is as much CO_2 as that which comes from burning. The products of life in these examples are not just *indistinguishable* — these are *the same* as the ones that are not products of life.

The latter implies something important. Because if the human-made artefacts are mostly well distinguishable from the things that are naturally formed in the non-living world, then due to the biosemiotically well-argued shift of semiotic threshold from the border of culture to the border of life, the distinction between the natural and life-made becomes structurally indistinguishable. In other words, what is *made* turns out to be both identical and non-identical to the things what are *not made*.

This is a very interesting case from the logical point of view, because on the one hand the distinction would need a term, but on the other hand, if we would introduce such a term, this would inevitably lead to a wish to define the qualitative difference — which is absent. Life is qualitatively different from non-life, but what it produces is both different and non-different.

How to solve this problem?

The solution would obviously require a more detailed description of the semiotic control.

The functional cycle (in Uexküll's sense) as a model of semiosis can be of some use here. It has always a double relation (recognition and action) to the object. This demonstrates well that from the side of recognition, the distinction is always qualitative, because the recognition of an object is controlled by memory. The results of an action (production, manufacturing), however, are not directly controlled — the only way to control it will be again via a recognition. The action does some work — but this work may do almost everything. It may build, and it may destroy. In this sense, as the activities of life, even building and destroying turn out not to be basically different. Decomposition and synthesis are equally the parts of life's metabolism and activities, and these may become distinguishable only for some higher forms of life; both may need energy, both may need semiotic control; there is no principal difference at the level of enzymatic processes, whether the process is establishing or removing a chemical bond. Both may be exergonic or endergonic. Even the concept of negentropy does not make a difference here. Thus, indeed, life (the semiotic control) may influence almost any process in almost any way.

Which means that knowing obviously always does more than it knows.

H.P. You are right. There is no simple way to distinguish a molecule that is synthesized under semiotic control from exactly the same molecule arising spontaneously. I discussed this problem in a paper titled "How does a molecule become a message" in which I concluded:

> A molecule does not become a message because of any particular shape or structure or behavior of the molecule. A molecule becomes a message only in the context of a larger system of physical constraints which I have called a 'language' in analogy to our normal usage of the concept of message. (Pattee 1969)

And as we agree, the simplest language or semiotic control process arises in the simplest self-replicating unit.

The principles and discoveries

K.K. For me, there are two fundamental observations or discoveries — or results — upon which the whole semiotic biology stands.

The first is the explanation for the biodiversity of species, and the variety of the types of categorizations. This is the answer to the question "Why there are species in the living world?" To put it very briefly, the biosemiotic answer is that communication (biparental reproduction being a kind of communication) in the non-categorized set of individuals would not be stable (Kull 1992). In other words, this is to explain why communication creates discretizations.

The second is the plurality of objects in the semiosphere. A thing in the physical world is just one, whereas in the semiotic world it is always many, it just cannot be one until it has a meaning (Kull 2007). Semiosis makes the world plural. Like, for instance, a painting — physically, it is a concrete pattern of pigments, but semiotically it is many things that can be recognized (or to what it refers).

From your point of view, what are the most important observations that motivated your interest in biosemiotics? And what are the important biosemiotic discoveries?

H.P. Living systems have always been a challenge, even a threat, to physicists who believe their laws are universal in principle, but appear to be no help in explaining life. *How do you explain why living systems are so clearly different from non-living systems when they both obey exactly the same laws?* That was the question that first motivated me. I first saw this question in Karl Pearson's *Grammar of Science* (1937: 287), a copy of which my Headmaster gave to me in the eighth grade. I still have the book and refer to it. Many physicists worried about this problem, like Erwin Schrödinger, Niels Bohr, and Max Delbrück, who are well known for their writing on the subject. Linguists, on the other hand, are understandably not concerned about this problem.

Roots and reminiscences

K.K. Semiotic biology is polyphyletic — it has several roots. Even the term "biosemiotics" has been coined independently a couple of times.[7] Contemporary biosemiotics includes scholars from different backgrounds and with different uses of terminology, who, after having recognized that their understandings match, accepted to build a shared conceptual apparatus, a common discourse.

The components of ideas that led me finally (at the end of the 1980s) to biosemiotics, as I would reconstruct them now, include seemingly (a) my early and strong (and continuous) interest in theoretical biology (which led me to exchange a few letters with Conrad H. Waddington and Robert Rosen, in the 1970s); (b) a strong semiotic (however, mainly cultural semiotic) school in Tartu; (c) contacts with biologists of non-neodarwinian views (on the one hand among Russian scholars, followers of the school of Lev Berg and Alexandr Lubischev, including the biologists of my own generation Sergey Chebanov and Alexei Sharov, and on the other hand the scholars carrying the tradition of Karl Ernst von Baer in Estonia); (d) my former participation in the research group of animal behavior studies where I came across Jakob von Uexküll's works (the search for traces of him resulted in contact with Thure von Uexküll, and via him with Thomas Sebeok); and certainly (e) the modelling research I carried out via which I understood the mechanism that is responsible for the emergence of species (which is very close to Hugh Paterson's recognition of the concept of species). After all this, and since the

---

[7] Friedrich Rothschild's (1962) and Juri Stepanov's (1971) usage of the term "biosemiotic" were probably independent. I have asked Juri Stepanov about this, and he said, "What concerns the term 'biosemiotics' — I had not heard it from anybody in 1971" (from J. S. Stepanov's letter to K. Kull, February 2010). T. Sebeok may have coined it independently and was using it in conversation before putting it in writing. The initial source of the same term for Marcel Florkin (1974) remains unclear.

meeting with our Danish colleagues Jesper Hoffmeyer and Claus Emmeche in the early 1990s, biosemiotics remained the stable name for the work that followed.

Histories of life, of course, are always plural. What are the paths that led you to biosemiotics?

H.P. Well, it was not only Pearson's question of why life is so different from non-living systems when they both obey exactly the same laws. It was Pearson's idealistic view even about physical theory that replaced my naive realism in thinking about both physics and biology. He made me see how all of our models are based on epistemological assumptions and limited by our modes of thought. Einstein's epistemology was influenced by Pearson's *Grammar*. Heinrich Hertz expressed these limitations of our models in his *Principles of Mechanics* (Hertz 1956 [1894]: 1–2):

> We form for ourselves images or symbols of external objects; and the form which we give them is such that the logically necessary consequents of the images in thought are always the images of the necessary consequents in nature of the things pictured.
> [...] For our purpose it is not necessary that they [images] should be in conformity with the things in any other respect whatever. As a matter of fact, we do not know, nor have we any means of knowing, whether our conceptions of things are in conformity with them in any other than this *one* fundamental respect.

Besides Hertz's separation of the knower and the known, there was von Neumann's (1955: 419–420) discussion of measurement in which he shows why an *epistemic cut* between them is a conceptual necessity, although its placement is largely arbitrary. It was also von Neumann's (1966) logic of self-replication that made clear the necessity of symbolic description as distinct from material dynamics to allow an unlimited evolution of novelty.

I have acknowledged elsewhere (Pattee 2001) some of the other physicists, biologists, and philosophers that have influenced my thought.

K.K. There are several approaches and scholars whom we can identify as biosemioticians but who themselves did not know or use that term. For instance, after reading Robert Rosen's (1991) *Life Itself*, I realized that he had reached the biosemiotic understanding — his emphasis is on the triadic relation.

H.P. There were indeed many physicists and biologists who, beginning in the 1950s, belonged to what Gunther Stent called the Information School of molecular biology (Stent 1968). It was generally recognized by this group that there was more to biology than just the molecular structures of DNA and proteins. Their focus on information was clearly a semiotic perspective.

Rosen was not a part of this group, but his emphasis on "relational biology" depended on semiotic rather than material relations. Rosen and I were friends for many years, beginning with our studies of hierarchy theory in the 1960s. To us, hierarchies, like measurement, are also dependent on semiotic distinctions because hierarchical levels are recognized by the necessity of different descriptions. We also focussed on epistemology. Rosen's modelling relation was based on Hertz's statement above (Pattee 2007).

The way to proceed

K.K. A large part of the existing biosemiotic studies has been devoted to theoretical and philosophical questions. However, what should the semiotic approach mean for biological experiments and observations, what is its relationship to empirical studies?[8]

---

[8] Some points on the role of biosemiotic empirical research are described in Kull, Emmeche, Favareau (2008).

H.P. I see this question as the central issue for biosemiotics. Earlier in our discussion you mentioned the "need for a special term" for structures arising from semiosis. This terminology problem is a symptom of a larger problem that biosemiotics is facing. It is already clear from our discussion that my physics language is different from your semiotic language; but the problem is deeper than language. Physics and semiotics have two very different cultures, and biochemistry is a third culture. The problem is even worse because all these areas have subcultures with their special foci and terminologies.

I'm sure you are aware of this culture problem. The two of us are both motivated to try to resolve our different language problem by discussions like this one. Unfortunately this is not the common motivation of most biochemists. When they are confronted with the biosemiotics perspective, they often resist semiotic expression of the problems of life as nothing but restatements of what they describe in their well-developed material language, which they regard as a more scientific description of life.

It is not clear to me what biosemiotics wants to be. All I can suggest is that if its practitioners want it to be accepted as science rather than as philosophy, they must focus more on empirically decidable models, rather than emphasizing its linguistic and philosophical foundations. In other words, if biosemiotics claims that symbolic control is the distinguishing characteristic of life, and if it also claims to be a science, then it must clearly define symbols and codes in empirical scientific terms that are more familiar to physicists and molecular biologists.

On the other hand, if biosemiotics is not primarily the study of symbolic matter but the study of symbolic *meaning*, then as I have emphasized (Pattee 2008), this requires a different epistemological principle than does the study of physics and biology. It will also require a language more familiar to philosophers and linguists.

One must keep in mind that the biosemiotic concepts like symbol function and meaning arise only by natural or cultural *selection* from those constraint structures that *physical laws do not determine*;

and yet all physical laws as well as all scientific models must be expressed in such symbol systems.

K.K. What should be the main biosemiotic questions on which the further research in biosemiotics should focus? Can we give a brief list of these?

H.P. Again, it is not clear what the main contributions of biosemiotics will turn out to be. As we learn more about the complexity of genetic expression, the analogies of genetic memory and natural language may not carry beyond the fact that they both use discrete, linear strings from a small arbitrary alphabet. So far, we have found nothing in the network of neurons in the brain that interprets sentences anything like how the cell interprets genes by the construction of proteins.

Molecular biology is currently totally involved with sophisticated technologies trying to unravel the functions and linkages in the masses of gene sequences data. These technologies already have specialized names like genomics, proteomics, and even transcriptomics. Even though all these studies could be correctly described as biosemiotics, I think it is very unlikely that the biosemiotic literature will alter the style or language of these highly competitive and incredibly complex empirical technologies.

In my opinion, biosemiotics will make the most lasting contribution by addressing the classical problems inherent in symbolic description and control of material systems at all levels — *the symbol–matter problem*. In this way it will contribute most to the epistemic foundations of all the sciences, of both the living and the non-living.

K.K. The main reason why we are developing the biosemiotic concepts is obviously just our wish to understand why and how life works. Since the questions we are dealing with are quite fundamental and related to several central questions of biology, it will also mean a reformulation (or rebuilding) of theoretical biology in many of its parts. Much of it comes out as a consequence from the application of the models of semiosis. The biosemiotic improvement of models of semiosis would probably also influence the whole theory of semiotics, which in its turn has consequences for humanities and for the relationship between physical sciences and humanities.

References

Eddington, Arthur Stanley 1929. *The Nature of the Physical World*. Cambridge: Cambridge University Press.

Elsasser, Walter M. 1998 [1987]. *Reflections on a Theory of Organisms: Holism in Biology*. Baltimore: Johns Hopkins University Press.

Florkin, Marcel 1974. Concepts of molecular biosemiotics and of molecular evolution. *Comprehensive Biochemistry* 29A: 1–124.

Hertz, Heinrich 1956 [1894]. *The Principles of Mechanics Presented in a New Form*. Mineola: Dover Publications. [Jones, D. E.; Walley, J. T., trans.; Original German edition: 1894, *Die Principien der Mechanik in neuem Zusammenhange dargestellt*, Leipzig.]

Hoffmeyer, Jesper 2008. *Biosemiotics: An Examination into the Signs of Life and the Life of Signs*. Scranton: Scranton University Press.

Jablonka, Eva; Lamb, Marion J. 2005. *Evolution in Four Dimensions: Genetic, Epigenetic, Behavioral, and Symbolic Variation in the History of Life*. Cambridge: MIT Press.

Kull, Kalevi 1992. Evolution and semiotics. In: Sebeok, Thomas A.; Umiker-Sebeok, Jean (eds.), *Biosemiotics: Semiotic Web 1991*. Berlin: Mouton de Gruyter, 221–233.

Kull, Kalevi 2007. Biosemiotics and biophysics — the fundamental approaches to the study of life. In: Barbieri, Marcello (ed.), *Introduction to Biosemiotics: The New Biological Synthesis*. Berlin: Springer, 167–177.

Kull, Kalevi; Emmeche, Claus; Favareau, Donald 2008. Biosemiotic questions. *Biosemiotics* 1(1): 41–55.

Neumann, John von 1955 [1932]. *Mathematical Foundations of Quantum Mechanics*. Princeton: Princeton University Press. [Beyer, Robert T., trans.]

Neumann, John von 1966. *Theory of Self-Reproducing Automata*. Urbana: University of Illinois Press. [Burks, Arthur W., ed.]

Pattee, Howard H. 1969. How does a molecule become a message? *Developmental Biology* Supplement 3: 1–16.

Pattee, Howard H. 1971a. Physical theories of biological coordination. *Quarterly Reviews of Biophysics* 4(2/3): 255–276.

Pattee, Howard H. 1971b. Can life explain quantum mechanics? In: Bastin, Ted (ed.), *Quantum Theory and Beyond — Essays and Discussions Arising from a Colloquium*. London: Cambridge University Press, 307–319.

Pattee, Howard H. 1988. Simulations, realizations, and theories of life. In: Langton, Christopher G. (ed.), *Artificial Life*. Reading: Addison-Wesley, 63–77.
Pattee, Howard H. 2001. The physics of symbols: Bridging the epistemic cut. *BioSystems* 60: 5–21.
Pattee, Howard H. 2007. Laws, constraints, and the modeling relation — history and interpretations. *Chemistry and Biodiversity* 4: 2272–2295.
Pattee, Howard H. 2008. Physical and functional conditions for symbols, codes, and languages. *Biosemiotics* 1(2): 147–168.
Pearson, Karl 1937 [1892]. *The Grammar of Science*. London: J. M. Dent and Sons.
Polanyi, Michael 1968. Life's irreducible structure. *Science* 160: 1308–1312.
Rosen, Robert 1991. *Life Itself*. New York: Columbia University Press.
Rothschild, Friedrich S. 1962. Laws of symbolic mediation in the dynamics of self and personality. *Annals of New York Academy of Sciences* 96: 774–784.
Stent, Gunther 1968. That was the molecular biology that was. *Science* 169: 390–395.
Stepanov, Juri S. 1971. *Semiotika*. Moskva: Nauka.

Chapter 13
A Roundtable on (Mis)Understanding of Biosemiotics

Claus Emmeche, Jesper Hoffmeyer, Kalevi Kull, Anton Markoš, Frederik Stjernfelt, Donald Favareau

On June 11–14, 2007, the International Association for Semiotic Studies convened its Ninth World Congress, at the University of Helsinki. The theme of the conference was "Communication: Understanding and Misunderstanding". In keeping with the conference theme, a roundtable panel discussion entitled "Understanding and Misunderstanding the Interdiscipline of Biosemiotics" was presented, wherein six of the founders of the contemporary project of biosemiotics attempted to explicate for those in attendance not only what the "biosemiotic project" of scientifically examining natural sign relations entails — but also to clarify the many misconceptions about biosemiotics that they so often find themselves having to explain both to other biologists, as well as to other semioticians. What follows is a transcript of that discussion.

D.F. Let me begin by explaining the purposes of this roundtable. It is now almost 23 years since Tom Sebeok made his proposal for the development of a semiotics — and for a view of life sciences — whose goal would be to transcend the explanatory limitations of both naive realism and naive idealism, so as to let us better examine and understand what we see happening in, and as, living organisms.[1]

---
[1] Anderson *et al.* (1984).

Most of the guests in our roundtable today trace their participation in Sebeok's project of biosemiotics back to its earliest days in the late 1980s[2] and have remained active in that project ever since. Our goal here today, in keeping with our conference theme, is to see if we can clear up a few "understandings and misunderstandings" about biosemiotics for the people who are with us here today — and also to get a kind of historical "snapshot" or record of where the project of biosemiotics happens to be at this point in time.

So I'm not going to give a long introduction to all these people. Rather, I'm going to ask them to say a few words about themselves. Starting from the far left, we have Jesper Hoffmeyer from the University of Copenhagen (*J.H.*); Frederik Stjernfelt, also from the University of Copenhagen[3] (*F.S.*); Kalevi Kull, Tartu University of Estonia (*K.K.*); Claus Emmeche, Niels Bohr Institute of Copenhagen (*C.E.*); and Anton Markoš, Charles University in Prague (*A.M.*). I'm Don Favareau, from the National University of Singapore (*D.F.*).

All right, then, let's start like this: No one here was trained as a "biosemiotician". So, prior to your involvement with this project, what were you doing? How did you come around to doing biosemiotics and what is your purpose in doing it now?

J.H. Well, I'm a biochemist by training, and I'm still a part of the Molecular Biology Department in the University of Copenhagen. But at a stage very early on in my career, I wondered why it was that my colleagues would go around saying such a lot of stupid things as "people are nothing but their genes" and all this kind of thing.

And yet, when I was young, I, too, was a believer in this way of seeing natural science. And I wasn't, at that early time in my intellectual development, opposed to the view that, in the end, human

---

[2]For instance, contributing to the landmark book *Biosemiotics*, edited by Thomas Sebeok and Jean Umiker-Sebeok — Emmeche (1992), Hoffmeyer (1992), Kull (1992), Stjernfelt (1992).

[3]Since 2009, University of Aarhus, Denmark.

beings were nothing but calcium ions and things like that. But the more I began examining the matter, the more I discovered that this cannot, in fact, be true. And so I started trying to think more deeply about such matters, just at around the time that my fellow scientists — and people in general — started talking about a new "Information Age" that we were entering.

And yet one of the things that became most clear to me was that, in the discourse of biology, one has a deeply incoherent notion of "information" being used all the time, where nobody really knows or attempts to explicitly define what exactly the term "information" means. Francis Crick, for example, when he defined the Central Dogma insisted that: "Information is something which goes always from the DNA to the RNA to the proteins. And never the other way." But what is this "something" that Crick is calling information? We know that it is not the individual DNA nucleotides themselves, for these give rise only to the aptly-named transcription and translation products which are not themselves those nucleotides, nor copies of them. Under Crick's formulation, then, this "something" that is genetic information seems extremely like a "cause" in the sense of Aristotelian efficient causation. But if you have discovered such a "cause" why then talk about "information?"

And eventually I came to realize that there is, indeed, a fundamental reason why it only makes sense to think about such phenomena as "information" in the semiotic sense and not as "cause" in the materially efficient sense — though, obviously, material causation and semiotic causation are always found together and stand in critically important relations to one another in any living system. Therefore, as a biologist, it appeared to me that a very deeply informed concept of "information" would become central to the study of life processes. And, indeed, the word "information" at any rate, did become very central to biology, in those years. But, as I say, the necessary explanatory concepts behind the promiscuous kind of "placeholder" use of this word were lacking. So Claus and I started wondering what exactly information means in living systems, and well, maybe Claus would like to say a word or two about this here, as we collaborated together a lot in the early days of biosemiotics.

C.E. Well, yes, I too started out as a biologist, doing my PhD work with Jesper as my supervisor and this was a project to articulate what biological information really is — because biologists, of course, are talking in "informational" terms all the time — though, as Jesper mentioned, without providing any really solid definition of just what "information" means in a biological context. So ours was an attempt to do theoretical biology on a more reflective basis. And in this project, we came across the work of Sebeok and others who had speculated profoundly upon the interface between biology and semiotics. So we thought that we'd, too, do well to learn some Peirce and to learn something about semiotics. So that was how I transformed from a theoretical biologist into a quasi-, if not yet full-blown, biosemiotician.

J.H. So I think it's important for you to understand and to see that the reason why Claus and I went into this and tried to get at the semiotics of life, was that we needed it for biology. I never was very interested in philosophy, or semiotics for that matter, but I had to do these things in order to understand biology. So I don't come here today in order to understand philosophy or semiotics. I'm here because this is the only way for me to reframe biology so as to make it understandable.

D.F. What aspects of biology couldn't you understand by using the explanatory tools that biology already has?

J.H. I would say that it's the "cumulate". For the way that biological molecules are organized, the way the cells are patterned, the way that everything is working together in the body is — as I see it now — determined by a semiotic kind of logic. And this logic simply cannot be derived from physicalist–reductionist science as it is presently practiced. You must add to that science an investigation into the kinds of semiotic organizing principles that allow the body and the cells — as well as biological eco-systems, for that matter — to actually do their work.

D.F. All right, well, I'm sure that we'll hear more on this, but just to complete the Copenhagen part of the story — Frederik, you did not come from a biology background in the same way that Claus and Jesper did. What is your interest in this project of biosemiotics?

F.S. Well, my story is a sort of mirror version of the one that Jesper and Claus just gave. I recall in one of our first conferences, Tom Sebeok came up to me and asked: "Are you a biosemiotician or are you a normal semiotician?" I thought a bit and answered, "I am a normal semiotician." So I think I'm not quite a biosemiotician to the same extent as Claus and Jesper. But in some way I turned out to be a sort of "fellow traveler" of these two and some others.

And this was for the reason that my background is in comparative literature — and in fact, I still have a position in that department at the University of Copenhagen. And my motivation was, again, a sort of mirror inspiration of the kind that we just heard about from Claus and Jesper, regarding their frustrations at how things were being thought about in biology at that time. Because I was similarly becoming desperate at the emptiness and evasiveness of so-called post-structuralism, deconstruction, radical social constructivism, and all this rubbish — to put it briefly — that was going on in my own area at that time.

And I wanted to move in a direction of a more realist semiotics — one where one could claim more than just to say that "there is nothing outside the text" or that every signifier begets another signifier, and so on and so on, *ad nauseum*. And this was why I've turned to biosemiotics, and that is how and why I became a sort of a biosemiotic fellow traveler.

D.F. Good. I hope that you will all try to help us to address the question of the difference between mainstream science and biosemiotic science, and mainstream semiotics and biosemiotic semiotics as we continue…but let's turn now to Kalevi Kull. Rather than coming from a background of comparative literature or molecular biology, you started your career as a botanist, is that correct?

K.K. I have always been interested in the best kind of biology. A botanist and a biologist, yes, also an ecophysiologist and theoretical biologist. But I see myself as a very slow person, so that is why I had to begin my investigations very early. So even back when I was in pro-gymnasium, I found myself beginning to ask questions about the nature of this endeavor called "the science of biology" and by corresponding with my friend Sergey Chebanov (a biologist who later became a biosemiotician), I gained much knowledge and insight from him.

Later on in my career, feeling myself very much to be a naturalist and a field biologist too, and well-embedded already in the society of professional biologists, I came to understand that the nature of life — as well as the investigation into the nature of life — is something that needs very much kindness and care. And this led me to search out those sub-societies of scientists in biology who would approach both living and the investigation of the living in this way. And so of course I found these people here!

Still, this search took a pretty long time. Because it was exactly 30 years ago, in 1977, when we organized in Puhtu, Estonia — the place where Jakob von Uexküll wrote his *Bedeutungslehre* [Theory of Meaning] — our Spring School conference "Towards a Theory of Organism" dedicated to Jakob von Uexküll, and this included a paper sent especially to that symposium by Robert Rosen.[4] One year later, in 1978, we organized with some Russian friends a conference entitled "Biology and Linguistics" in Tartu. After that, however, it still took 15 more years to get other people to see the light that Thomas Sebeok and Jesper Hoffmeyer were already seeing at this time.[5]

---

[4] Published later as Rosen (1984).
[5] See Kull (1999).

A Roundtable on (Mis)Understanding of Biosemiotics 241

D.F. Anton Markoš, you are both a philosopher and a biologist, is that correct?

A.M. Well, I am sitting in the department of Philosophy and History of Sciences in the faculty of Sciences at Charles University in Prague. So as you can imagine, the floor is sometimes quite hot for us there…but we like it!

And yes, by training I am a biologist — and when someone chooses to study biology, it means that one is eventually going to have to confront the question how living forms came into existence. But today, when you take a course on evolution, and you want to know how bodily forms came into existence, you learn in that course that the definition of "evolution" is "change in the frequency of alleles in a population over time". And with this, one suddenly learns that biology has become very "scholastic" in its arena of inquiry and in its definitions. Because with such "virtual" definitions, you can do a very disembodied kind of population science, you can do statistical analysis and calculus — but you find out that, all of a sudden, you no longer can account in a scientific way for actual living bodies and their actual living experience. Therefore, despite any reasonable scientific inquirer's demands of *habeas corpus ad subjiciendum*, one soon finds out that there is no longer any room for this full sense of "body" in, of all places, biology!

Realizing this, I then moved over towards the philosophers. But to my astonishment, I found out that the situation is much the same there! So everything — all the sciences and the humanities — have become unworkably scholastic, allowing no space for the actual workings of life. Because if life means bodily existence, and if bodily existence entails and is made possible by biological reproduction; then it follows that this reproduction cannot be mere copying of genetic alleles because in the bodily world, one thing cannot copy any other thing. At best, the living body can only re-produce forms to the extent that one can that make structures that are similar to pre-existing ones, but not the same. Yet in order to make similar structures, the body must somehow recognize likeness — otherwise it cannot determine in what sense any two things are similar. And in order to recognize likeness, it needs experience.

So if you take all this: likeness, experience, maybe memory, and body; you'll find there is no room for all of it in either of the two realms — philosophy and science — that are connected in our department — though without doubt, all of this is deeply connected in life. Realizing this, I started looking around to see what community, if any, could offer some help with these issues. Well, I found these people, and, so far, I think that the [Peircean] concept of "abduction" has turned out to be the most helpful idea that I have found during my searches.

D.F. Thanks, everyone, for those brief self-introductions. Now just to stay at a very basic level here — please, be clear to us: Why do we need a study of "semiotics" undertaken from within biology? Or to ask the question a little differently: What is it exactly that you are trying to "add" to our understanding of the biological world?

F.S. For an answer to this, I would go back to Jesper and Claus' now old, but still ground-breaking 1991 paper,[6] where they laid out a very basic argument which is still for me the basic argument for the coming into existence of biosemiotics. Namely the fact that if you take any textbook in biology — and you can make a test of this for yourself — anywhere from biochemistry and molecular biology and all the way up to ecology and ethology, all the branches of biology; if you take an ordinary textbook, and pick some arbitrary page, you'll find semiotic terminology. You'll find biologists talking about signs and codes and communication and information — and the strange thing about it is that these concepts in ordinary textbooks do not appear as technical concepts. They appear as straight-off, common sense concepts.

And in many cases, if you go up to the person that wrote that textbook and ask him: "What do you mean by 'code' and

---

[6]Hoffmeyer, Emmeche (1991).

A Roundtable on (Mis)Understanding of Biosemiotics

'communication' and all this business about intercellular 'messaging' and so on?" Then he will answer, "Well that's just a sort of metaphor. Just a sort of image. If I tried, I could explain the same thing without any semiotic terminology." And then the obvious counter-question to that is: "Then why don't you do it?"

It is obvious, then, that biology is unable to get rid of these concepts — and so the next step that biosemiotics proposes is to say: "But then why not try to use those concepts in a more coherent, more scientific way where you technically define them? So we can agree upon what we mean when we say 'code', 'symbol', 'communication' etc. etc." So that these terms stop being vague, common sense notions are qualified as real technical terms of science. That is still for me the most basic argument for why we need a biosemiotics: because the ubiquity of semiotic processes are already described in every biology textbook — but only in a severely undeveloped, at best possibly embryonic, conceptual form.

D.F. It seems to me what you're saying is that we've already cashed out the scientist's view of "message" or "communication" — and that this keeps coming back down to merely the quantified descriptions of the chemical substrate whereby that message is realized. But you want to talk about the message qua message, right? The sign as a sign, and not just as its material substrate, is that correct?

F.S. Yes.

D.F. Well how would one go about trying to do this? There must be many ways that one could attempt to undertake this kind of investigation, I would imagine. Are there some approaches that you have decided are more useful than others here?

J.H. Well, as I mentioned earlier, the whole notion of "information" has proved to be incoherent in the study of biology, if you attempt to treat information as a kind of "entity" or "cause" — as most biologists unreflexively still tend to do. And yet as soon as one applies a triadic conception of the sign to those cases where one talks about "information", it quite quickly becomes clear that this kind of an understanding does, in fact, help us to better understand exactly what it is that we are trying to explain about the processural reality of biological "information" to an organism.

And in this way, one can escape the reductionism that is implicit in the debased notion of "information" that is currently being used in biology. Because thinking of information as if it was a "cause" will inexorably guide you into a reductionist view. And the way to escape this kind of explanatory dead-end is to expand one's conception so to see that "information" is only information in the sense that it is something that has to be interpreted.

Understanding this, if one then asks: "What is the unit that interprets?" the answer is that, at the most fundamental level, such bodily interpretation must take place at the interface of all biological sign processing, that is, the living cell. This kind of bodily — and not yet mental — interpretation takes place simultaneously on many different levels; these levels include, of course, the entire body system proper, but also the level of the subsystems, such as the nervous and the immune systems, as well as the sensory apparatuses, such as the surfaces of the eye or skin. And as a result, you have interpretants feeding into interpretants upon interpretants in one long chain — or web. And this is what "knowing" ultimately is. So in this way, the formerly mysterious notion of biological "information" finally does begin to make sense when you start looking at it from a biosemiotic, rather than a materialistic or a psychological, perspective.

D.F. And you are saying that if mainstream biology adopts this perspective, it would have a fuller understanding of the sign-processing phenomena that it has heretofore been looking at. I guess the obvious question to ask then, is: "Would science then have to throw out all of the findings about the material substrates of these interactions that it's already achieved?"

J.H. Not at all. I mean, all of us here, as scientists, are very grateful for the hundreds of years of experimentation into the physics and the chemistry of life, and for all the things that doing science this way has allowed us to find out. But just the physics and the chemistry alone doesn't fit together in a way that can fully explain biological life. And to do this, one has to reframe what's known about physics in light of what's known about life. That's what biosemiotics is: a reframing of the kinds of things that a responsible biology is going to have to look at in order to scientifically explain the processes unique to living systems.

D.F. And how successful has this project been? To get scientists to see things this way?

C.E. I find two main reactions that one often gets from scientists. One reaction has already been phrased here as a question, and that is: "Why must we redescribe what we already know from standard science?" And I think one answer to that is that standard molecular biology and cell biology and certainly genetics, have been quite reductionist in their approach. But even in those fields today, there is now quite a bit of recognition now that such reductionist approaches are not really working, and that what they need are more synthetic and holistic models. There is, therefore, on the one hand, an acknowledged need in science to responsibly join together all of their disparate data within a wider explanatory perspective. Biosemiotics is also a kind of "widening", then, of science, so as to accommodate a greater dialogue between its fields. And in a complimentary fashion, it is also always helpful intellectually to provide several distinct theoretical models in the attempt to explain the same set of basic factual findings — "your data", if you like. Because such distinct models can provide really new perspectives, and therefore new theoretical understanding, about what it is that you are seeing.

For example, at the annual *Gatherings in Biosemiotics* conference that we had in Groningen just last week,[7] there were several papers that gave very good explanations from a semiotic perspective on things that you couldn't really get the same understanding of if you

---

[7]See Neuman 2007.

just stayed within the mechanistic paradigm of, say, molecular biology. And you also see in the fields of biosemiotics a diversity of interests: some people are doing work on cellular signaling and genetics, others are interested in animal behavior, others in the organization of human sign use, and others are interested in its more broader perspectives — its metaphysical and philosophical implications.

And in this sense, I think that biosemiotics is very much like cognitive science. There, too, you have an interdisciplinary field, and you see a lot of philosophers interested in metaphysical issues — but you also see practicing scientists who are much more narrowly dedicated to solving particular scientific issues. So biosemiotics and cognitive science have some similarities in this respect. They're both cross-disciplinary. They are both dealing with cognitive informational processes. But I would say that the "information" metaphor in cognitive science is more mechanistic and less thought-out than the "sign action" concept in biosemiotics.

D.F. I suppose that a scientist coming to biosemiotics for the first time would ask: "Why is there all this emphasis in biosemiotics on umwelt, and on subjective experience? Isn't this exactly what science wants to get away from — subjective explanations of all sorts? So what is the purpose of talking about the animal's subjective experience? What's scientific about that? Or is this just a kind of storytelling?"

K.K. There was a very nice conversation published between Howard Pattee and Robert Rosen — and I would include both of them in biosemiotics, I think they are both biosemiotic thinkers — and they were asked: "What exactly is the aim of biologists in biology as they see it?" And Rosen gave a very clear answer: "What is important in biology is not how we see the systems which are interacting, but how they see each other".[8]

D.F. And this includes animals.

---

[8] Rosen, Pattee, Somorjai (1979: 87).

K.K. Yes, animals, and all other organisms. Now — this is an almost impossible task for a purely physicalist science — I would say, to show and to discover how organisms see the world; what is there in their umwelten; what do they know. This demand requires us to carefully develop a new set of methods, and a new set of conceptual apparatuses, in order to somehow grasp all this.

J.H. And so, of course, one can either choose to continue to dance around all this, or to finally try to come to terms with this, head on. Now, if we don't think, as some philosophers do, that our own lived subjectivity or subject-hood or first-person experience — the feeling of an "I" — if we don't want to just write that off as an illusion, if we think that this experience, at any rate, is "real" — then we must explain within the context of natural evolution just how such a thing did, in fact, come about. And while I suppose that it is still conceivable that the information theoretic or the computational approach may someday tell us something about how already living agents operate so as to build their intelligence, I don't think that's the hard problem. The hard problem is the evolution of the experiential world itself — agency, if you like, and intentionality.[9]

Now evolutionary theory in biology must explain how it is that we are not miracles, that we did not come down from heaven, but that we are somehow born out of nature. Thus, we must provide an explanation from within evolution that accounts for the experience of intentionality and subjectivity in living creatures. We can't go on just denying that it exists, because it's such a large part of our own life — and, in fact, the life of every animal. It will do no good to write off this fact of biology as an "illusion". So for that reason, I believe that we have to go back and see if we can make a biological, and evolutionary explanation for how subject-ness, or agency and intentionality, has come about.

D.F. And what is the role of "semiosis" in this new evolutionary explanation?

---

[9]Hoffmeyer, Kull (2003).

J.H. A fundamental one. Because semiosis is the relation, in action, of a sign and its object through the formation of an interpretant. This, of course, is a kind of semiotic causation, and thus a kind of causation. But it's a level of causation that does not reduce to simple efficient causation.

D.F. So in an evolutionary context, "semiosis" would establish the relation between an organism, its experience of the world, and the world itself, yes?

J.H. Precisely. Because those relations are exactly the ones that we are looking at — or should be looking at — when we are attempting to understand organisms' natural "evolution".

F.S. And in this context, I think that the strength of the sign concept is that it comes without any precondition of human-style "consciousness". Therefore we are able to bracket that issue and to focus instead upon the very structure of the sign relations that organisms actually use. But doing so also points out to us to a very important caution that I think most of us realize that biosemiotics must take. Most of us agree that it's an important task to try to construct an understanding of evolution which makes it understandable how we ourselves, with our human-style "consciousness", are also products of this process.

And to do this entails that we should be very cautious not to take the concepts that we use to conceptualize the evolutionarily latest and most complicated product of such semiotic processes — which is human culture — and uncritically project such concepts downwards. So when we employ all the many different types of sign concepts that we have in semiotics, we must always be very careful to distinguish which of these sign concepts are appropriate to the precise level of investigation that we are presently undertaking — and we must do this so that we do not commit the grave mistakes of vitalism or anthropomorphism.

D.F. Yet it is exactly the charges of "vitalism" and "anthropomorphism" that are often made against biosemiotics by those who may be hearing about it for the first time, but who are themselves insufficiently familiar with the actual literature, yes?

A.M. Yes! And to expand a bit upon Jesper's statement that when you open any biological textbook you'll find it said in the first chapter that "a living body is composed of…these and those chemicals…and thus it is composed of cells…it is composed of organs…" in the same way that a building is [passively] composed of bricks. And yet we all know unmistakably that, much unlike a building, and unlike any non-living system, a living body actively composes itself. It is that which it composes, and is so continuously, for as long as it is alive. So "a living body is composed of…" is a strange kind of "creationalized" statement, though almost nobody realizes it. It is not a reversion to vitalism, I would say, to quote French philosopher Raymond Ruyer's observation: "Gentlemen, we have never been dead!"[10] The point is, rather, that in living creatures, there has been a continuity in life-sustaining processes from the very beginning — a continuity unbroken now for four billion years. And it is in these creatures alone that we see experience, memory, forgetting — and in different lineages, different images of how the world is. Those are some very highly-qualified semiotics, I would say.[11]

D.F. So you're saying that at the heart of these things that are "organisms" today — and at the heart of human, and maybe even all animal experience and knowing — there lie simpler, prior relations that are necessary to organize the phenomena of life, yes?

A.M. Yes. We criticize reductionism, but it's not only reductionism. With genome projects, there are no reductions — one has a whole genome's vocabulary, and one can use that. And just by comparing genomes, one can make enormous systematics. But what is it for? For forensic medicine, for forensic evolution, and for making evolutionary trees. But it will not explain why I look like myself and

---

[10]Ruyer 1946, introduction.
[11]See, e.g., Markoš 2002; Markoš *et al.* 2009.

not like you. And why neither of us look like a horse. So if we don't accept this, and we don't want to say: "Well, our biology is good enough as it is", we have to stand up against that kind of thinking, small as we are. So I see us as advertisers of a new biology — but there are still one million old biologists standing against us! Our opposition is somewhat crazy, of course, because biology as it is is a very good science, and I say that as a biologist. But I think that two views of the same topic of "life" are better than one, as Bateson said. So why not develop a complementary view, wherein the goal is not to explain life as passively "composed of" but to explain instead how life actively "composes itself" — as shown by the phenomenon of evolution.

D.F. Would semiotics explain that? Would biosemiotics explain that?

A.M. Biohermeneutics would, I would say: for this is the bodily reading of the information written in our genomes, according to the experience of the evolutionary lineage that led to human beings.

D.F. So reading, interpretation, sign processes are critical to the organization of biology. And everyone here agrees with that.

A.M. Yes. And these topics have not been covered by standard biologists. So the question then becomes: "Are we going to change the paradigm of mainstream biology or will we make another competing realm?" Views inside the community of biosemioticians vary to a great extent — from "Let's create a hard science of biosemiotics" up to "Semiotics and science are incompatible". I cling to the second opinion: Biosemiotics should treat those aspects of life that biology — as a hard science — cannot rasp.

D.F. That's a good question. Are you going to make a competing realm, or what do you see happening?

C.E. Well, to address this question, let's consider a very common metaphor that often comes up when we discuss "What is the place of biosemiotics in science?" and this is the metaphor of "national territories". Very often, we tend to see each of the sciences and each of the scholarly investigations as somehow dividing the world up into separate territories, each with their special governments, boundaries, citizens, and laws. But maybe instead we should begin to think upon this investigative activity as more like a network, than as a series of disconnected and competing states. And I think that biosemiotics can help to establish better network connections between the so-called hard sciences and the humanities by not accepting this "national territory" metaphor of research. And in that sense, biosemiotics may be a central node, or at least one among several important nodes, in this whole hub of investigative activities.

J.H. Yes, and I see this very much as connected to a general trend in advanced science nowadays that goes beyond the deterministic thinking which characterized so much of the old science, where even quantum mechanics is essentially classical in this sense. But the new emphasis on self-organization, system dynamics, non-linear systems, chaos theory, and so forth — all of this points towards another kind of understanding, the understanding of emergence and emergent systems. Biosemiotics also is a way of figuring out how emergence occurs in the life sphere. Because organic "emergence" seems to be very much a thing that we have to take seriously now. So I am thinking that in one way, biosemiotics is already quite advanced science.

D.F. Agreed. But so far what everyone has been talking about today has been concerning a paradigm shift — or at least, the attempt to establish a different perspective — in how we think about and do biology. But now, what about the other side of the biosemiotic project? What, if any, impact or challenges does biosemiotics pose to how one thinks about or does semiotic inquiry? What, if anything, does a biosemiotic perspective add to a non-biosemiotic perspective on say, culture and the humanities? Any thoughts on that?

F.S. As I mentioned earlier, one of my reasons for affiliating myself with this field was a disappointment with certain trends in general semiotics. Before biosemiotics, before the whole Peircean renaissance in semiotics during the last ten years or so, I think it's safe to say that the main current in semiotics was structuralism. Now, many good studies were made there, but there were some basic flaws in the assumptions of structuralist semiotics. They very often tended to claim that semiotics was a uniquely human endeavor; that it could not be applied outside of our species; and that, moreover, it tended to claim that these human signs were in all cases completely arbitrary. And this led into all kinds of relativism and skepticism and all sorts of blind alleys that you'll find if you go back to the literature of 20 years back.

This is why I think that biosemiotics in some sense is a double-edged tool: it may help us to deepen our understanding of biology on the one side, but it may also help us improve our conceptual foundations for a general semiotics on the other side. And to re-establish semiotics in a more satisfyingly realist manner than the general semiotics that went before it.

K.K. I have to say, in addition, that there has recently been some very interesting work in biosemiotics that may prove impactful to more general and theoretical semiotics. Because up till now, most semiotic study has been limiting itself to dealing with human culture. But now we are witnessing an *ecological turn*, so to speak, taking place across many fields in the humanities. And because of this — *the ecological turn in semiotics* — the studies of cultural semiotics will need somehow to integrate the study of "natural" semiotics in their research, without reducing that semiotics to mechanics. Biosemiotics, of course, is the logical place to begin such an integration...as we've been doing all along!

D.F. An objection that is often made against biosemiotics is that talk about "human knowing" of the kind that might be studied in general semiotics, and talk about "non-human knowing" of the kind that might be studied in biosemiotics are insufficiently recognized to be talking about two distinct things. So the question I have for all of you is: "Are they, in fact, two distinct things? And if so, is there a principle distinction in biosemiotics that distinguishes human knowing from non-human knowing?"

F.S. Of course, being a Peircean, I would claim there's a continuum between them. But to claim that there's a continuum between any two things is not the same as to claim that they are equivalent. Because a continuum may take different trajectories and courses, and I think that everyone here agrees that the continuous evolution from higher primates into our species is one of the trajectories in evolution that is particularly steep. In that phase, due to the kinds of sign processing between individuals that became available, the evolution from primate to human acquired a very particular speed — and it is this speed that might make it appear that what we are looking at are two distinct worlds. But as far as the whole discussion about what the exact new sign-types appearing during that phase were — the question of the origin of so-called anthroposemiosis, or our contemporary style of human knowing — my personal guess is that the sign skill that we mastered, but that even our higher primate relatives did not, is what Peirce calls "hypostatic abstraction". That's my theory, anyway — but I could talk about that for hours, so I'd better stop now.

J.H. To address the same point, only using a technological metaphor: when the Industrial Age arrived, we didn't stop producing new grains. We still had to produce agricultural products. And when the so-called Information Age came in, we still needed to rely on mechanical energy for most of our production processes. And in a similar manner: when human consciousness started arising in the world, it's not that we stopped being animals. But certainly our ways of doing things became superseded by other means. In my own work, I refer to this concept as a process of "semiotic scaffolding". And we can see this taking place even on the most fundamental levels of the organism, as it is the way of building a structure in the organism by semiotic means that allows it to remain stable within its environment. And in fact, one can think of much of physiology as the semiotics scaffolding of body parts in relation to use in the world. And one can see that the DNA and the genes provide the semiotic — which is to say the sign-based — scaffolding of certain developmental pathways in ontogeny.

And in the same way, human culture, too, provides us with new species-specific semiotic scaffolding mechanisms. For culture itself is just an enormous, more productive way of scaffolding things. But I am quite sure that semiotically, we are still very much bodily scaffolded at the same time that we are culturally scaffolded. It is, of course, true that human understanding is considerably different and so much more semiotically complex than that of other animals. But nevertheless, the more animal kinds of understanding still seem to be very much alive in our bodies — and therefore still very much a part of human experience and understanding. And I would suggest that in the human sciences, one should be aware that one kind of understanding doesn't exclude the other. And I think that it would be a very interesting project — not just for biosemiotics, but for the humanities — to see how these scaffolded semiotic processes work together to make us the kinds of beings that we are.

A.M. I'm not so sure how many people are going to embrace that project, however. If you look at the last two hundred years of the history of biology, you can always find minor groups of dissidents claiming that "there is something more to life than is being accounted for in the standard science." You find this especially in the major works of literature. But who alone succeeded in actually establishing this case? Only one single person succeeded. It was Darwin. Darwin brought story back into biology, and into the sciences. And it took 150 years to push story back out of Darwinian evolution to make neo-Darwinism! Well, we've found the story in life again. But since we see that in 200 years, only one single person was successful in synthesizing history and working science, well…we must be careful in our optimism!

K.K. Yes, but "having semiotic eyes" allows us to read more books than do the neo-Darwinians! And it also gives us a good basis to see what is most reasonable from the other side. So we can see how narrow the view would be if our understanding of evolution was only that of gradualism. For we can see *why it must be so* that there are novelties, there are emergences, there are qualitative changes, there is punctualism — all which is so important to include in any honest explanation of evolutionary processes. And it is, in fact, *all* of this — evolution's qualitative and quantitative aspects both — that biosemiotics insists we must study and attempt to scientifically explain.

A.M. Of course, I agree. My point was only that most modern scientists are in many unseen ways "scientific creationists" — and not only because it took 150 years for science to properly acknowledge evolution. But because, in fact, there are only two histories, or "stories" that have been allowed into science. One is the Big Bang, where the whole story is effectively condensed into the first 10^{-42} seconds of the universe — after which you then just have normal functioning, the mechanical, clockwork functioning of the universe. So that story already finished long ago. But the second story — that of biological evolution — is still going on.[12] This is a problem that is posed before science all the time. And, of course, it is a challenge for us too.

D.F. I see that we are coming towards the end of our session. Before we go, though, let's take some questions from the audience for our panel.

AUD 1. A very concrete question. What are the possibilities for biosemiotics to become a practical, empirical research program? Is this allowed? Are concrete case studies needed?

---

[12] A very inspiring text in this respect is the recent book by Gilbert and Epel 2009; our own version of the story can be found in Markoš *et al.* 2009.

K.K. My view is that it should be done. Such studies can be very rich. But what is still needed is a well formulated semiotic methodology for biological empirical research — because such qualitative methods of study are not at all well developed anywhere in biology itself at this time.

J.H. I think my general answer to this question is that what biosemiotics can do, for now at least, is to direct the curiosity of scientists into still relatively unexplored areas, so that the kind of experiments that one does would be different. But I don't think, as such, biosemiotics is primarily a methodology for doing experiments.

D.F. Well, isn't it the case that if a biosemiotician runs an experiment, say, on a brain event, he's going to get the same results that a neurobiologist would, simply because the underlying biology is, of course, the same? Is it that only the two interpretations that the resulting data represent may be what's different?

K.K. No. Asking different questions may result in different findings.

J.H. And of course the aim would be that under the new system, you would have the conceptual means to see a logic in its working — and this might direct your investigations in another direction.

AUD 2. What about feedback? In a biosemiotic context, can, say, Wiener's concepts of "feedback" help build up the conceptual framework of biosemiotics today?

K.K. Yes, but it may not be as useful if one uses Norbert Wiener's mechanical model. But if you take Uexküll's functional cycle, this is also a feedback model, more closely designed from the observation of the organism's physiology, and this is a feed-forward model, in Robert Rosen's sense; this is one of the basic models that we have used.[13]

AUD 3. From what I have been hearing from some recent approaches to biosemiotic theory, I am quite skeptical that we can overcome

---

[13] Wiener (1948); Uexküll (1992 [1934]).

our anthropocentric access to experience in order to overcome our culturally shaped prejudices...

F.S. Okay, and that's a cultural prejudice that you are uttering — so now where are we?

AUD 3. True — my idea is that there is only the possibility of changing one set of cultural prejudices for another one. But the idea of umwelt requires you be able to get into the animal's mind and I cannot see how you believe that you can transcend your own biologically determined umwelt to do that. For even this making of a theory of biosemiotics is a very high level, culturally determined human activity. It requires a very long time of human schooling within purely cultural instructions in order to even be doing what you are doing now — that is, to express highly abstract ideas. So, while from that level of ethics and everyday care, we can very well overcome our anthropocentrism at the level of our contact with animals — people have done this, and have shown that this can be done. But this making of theory — this is a problem: how can we make general conceptual tools which are not a fruit of the cultural inheritance we all share?

D.F. I don't think that biosemiotics is trying to occupy some impossibly non-perspectival conceptual space beyond whatever we can achieve with a self-conscious understanding of our own anthroposemiosis. The project, I think, is not so much of trying to somehow experience the mind of the animal and then to write up a veridical report on that — rather, it is to try to investigate those self-world sign relationships that the animal sets up for itself that give rise to a network of actionable meanings in animal life. This network you could refer to as its "mindedness" if you prefer — but its investigation is asking a somewhat different question, and is amenable to different paths of inquiry, I think, other than trying to tell ourselves that we must be able to have the same experience that the animal — or even another person — is having before we can reasonably make any claims about it. That's just the kind of Cartesian dualism that we're trying to overcome.

F.S. I would also just say there is no general answer to your last question. I mean, the skepticism you are expressing might be well-placed in specific cases where you are going into specific interpretations of specific organisms. If you can point out anthropocentrisms there, I think you should do that. But I don't think there's any general answer to "skepticism" as a way of thinking.

AUD 4. Given my experience at the rest of this conference so far, I see that it is a problem inside the larger arena of semiotics, in that owing to the exclusive focus on human culture, other biological entities are de facto exempt from semiotic investigation. Perhaps biosemiotics is to be excepted here — yet I still see the potential for a problem here, too. For in doing biosemiotics, you are still using a "standard semiotic theory" — that of, for example, Peirce. But there are some terms that are critical to those theories, such as Peirce's notion of an "interpreting mind". And, of course, when you study a biological entity on the level of itself, the question remains: where is this mind? So it's a two-sided question. First, how will you circumvent the difficulties with already established semiotic terminology? And is biosemiotics liable to take semiotic theory further beyond just applying Peirce in biology?

C.E. When I first tried to apply Peirce, I found many different interpretations of his work that were essentially misinterpretations of Peirce — and having this idea that "the mind" is the name of some kind of mysterious entity that somehow imputes meaning upon the sign is a very common Cartesian misinterpretation of Peirce. Peirce, from what I can tell, would reject this kind of reified, non-processural view of "mind". So this makes relevant the whole question of the supposed "locus" of meaning: Is meaning the kind of thing that is physically "located" at a certain place? Is it — as people often say — in the mind? Or is it possible that talking about "the mind" is really not the right way to phrase the whole question of meaning? From going deeply into questions like these, I think that we can learn a lot. So I think it's very much an ongoing process of really understanding the best possible manner to interpret these questions of sign action — and by using Peircean terminology, we can hopefully non-naively ask such questions as: "What is the interpretant in process x? How many different kinds of interpretant signs do we have at our perceptual disposal? Or for use in our conceptual understanding?" And things like that.

F.S. There were two parts of your question. First, I think, that even if we are basically Enlightenment optimists, I am not convinced that we have yet made the huge kind of progress that we want to in biosemiotics. Up till now, for example, I don't think we have developed all the sufficient terminological units that we need in order to fruitfully talk. On one issue, however, I think we have been a success — and this is the one that you refer to in your first question. I think that in the general semiotic community, biosemiotics is much more accepted now than it was ten years ago. Ten years ago, all the old French structuralists were fighting to keep it out and I think that they have given up that fight now. So I think that within that small field, that's a partial success at least. And I do agree with Claus that if you take Peirce's concept of "mind" it has almost nothing to do with the common sense, ordinary, everyday-language notion of "mind". It's a much broader term and he underlines on many occasions that it does not imply that there needs to be any consciousness present, much less human linguistic activity. He merely defines "mind" as a certain sort of systematic processing of signs.

D.F. I see that we only have a few minutes left, so before we leave, I want to ask the panel: Regarding the state of Sebeok's "biosemiotic project" today, in 2007 how do you think it's going? Is it progressing? I know that it has often been said by biosemioticians that they are still doing the "backwoodman's work", as did Peirce, clearing the grounds of outdated ideas and marking the paths for others who will follow them. If that's so, where do you find yourselves today — still chipping away? Making inroads? Or do you see the project failing? What is your take on the state of the art in biosemiotics today?

C.E. Well, in principle, I'm an optimist, and I do see a lot of progress… But, of course, that depends upon how you measure "progress". I think that progress cannot best be measured by some of the ridiculous measures one sometimes see in standard science, such as "citation indices", or "the expansion of the number of workers in the field" and so on. I'm more concerned about progress regarding those big issues that you posed, such as the distinctions between human and non-human knowledge. Though I'm not sure that even biosemiotics in itself really can answer those big questions at this point. These questions may simply be too big for such a little bunch of workers.

J.H. Like Claus, I'm not sure how one would go about trying to chart biosemiotics' "progress", but what I am sure of is that somehow we are partners in lots of different approaches that are going on in science and in semiotics in these years, and that we are trying, in our own way, to make a new kind of framing of things. And for a long time I had the feeling that these ideas seemed to have fallen on barren earth somehow. I really have had that experience. But recently it seems that a lot of scientists are finally coming around and questioning the things that we have been questioning all this time. In fact, I was just in Boston two weeks ago at a high profile meeting with scientists like Stuart Kauffmann and Gregory Chaitin and several others, and I finally had the feeling there of biosemiotics being received. They really grasped it. This message was received. And I think that a lot of people are now thinking and working this way. So maybe biosemiotics as such will not build up into its own kind of separate field, but instead its ideas will profuse into scientific thinking. And I think that eventually, maybe, we here will be forgotten — but maybe we will have made a little bit of a contribution to science as a whole.

K.K. Yes. And, in any case, the concept of "progress" as having "high value" belongs to the Age of Modernity in science — and this is something that we have now left behind. Life leaves signs. But life also, and exclusively, *lives* in signs. Therefore, let us too live in signs, instead of just leaving them — because what has real value is to be, and to have been, *alive*.

D.F. Well, that's as "biosemiotic" a note as any to end on, I suppose. Thank you all for being with us here today.

References

Anderson, Myrdene; Deely, John; Krampen, Martin; Ransdell, Joseph; Sebeok, Thomas A.; Uexküll, Thure von 1984. A semiotic perspective on the sciences: Steps toward a new paradigm. *Semiotica* 52(1/2): 7–47.

Emmeche, Claus 1992. Modeling life: A note on the semiotic emergence and computation in artificial and natural living systems. In: Sebeok, Thomas A.; Umiker-Sebeok, Jean (eds.), *Biosemiotics: Semiotic Web 1991*. Berlin: Mouton de Gruyter, 77–99.

Gilbert, Scott F.; Epel, David 2009. *Ecological Developmental Biology: Integrating Epigenetics, Medicine, and Evolution*. Sunderland: Sinauer Associates.

Hoffmeyer, Jesper; Emmeche, Claus 1991. Code-duality and the semiotics of nature. In: Anderson, Myrdene; Merrell, Floyd (eds.), *On Semiotic Modeling*. Berlin: Mouton de Gruyter, 117–166.

Hoffmeyer, Jesper 1992. Some semiotic aspects of the psycho-physical relation: The endo-exosemiotic boundary. In: Sebeok, Thomas A.; Umiker-Sebeok, Jean (eds.), *Biosemiotics: Semiotic Web 1991*. Berlin: Mouton de Gruyter, 101–123.

Hoffmeyer, Jesper; Kull, Kalevi 2003. Baldwin and biosemiotics: What intelligence is for. In: Weber, Bruce H.; Depew, David J. (eds.), *Evolution and Learning: The Baldwin Effect Reconsidered*. Cambridge: MIT Press, 253–272.

Kull, Kalevi 1992. Evolution and semiotics. In: Sebeok, Thomas A.; Umiker-Sebeok, Jean (eds.), *Biosemiotics: Semiotic Web 1991*. Berlin: Mouton de Gruyter, 221–233.

Kull, Kalevi 1999. Biosemiotics in the twentieth century: A view from biology. *Semiotica* 127(1/4): 385–414.

Markoš, Anton 2002. *Readers of the Book of Life: Contextualizing Developmental Evolutionary Biology*. Oxford: Oxford University Press.

Markoš, Anton; Grygar, Filip; Hajnal, László; Kleisner, Karel; Kratochvíl, Zdeněk; Neubauer, Zdeněk 2009. *Life as Its Own Designer: Darwin's Origin and Western Thought*. Dordrecht: Springer.

Neuman, Yair 2007. The seventh Gathering in Biosemiotics: A review. *Sign Systems Studies* 35(1/2): 301–303.

Rosen, Robert 1984. Organismi mõistest. [On the concept of the organism.] *Year-book of the Estonian Naturalists' Society*, vol. 69 (Problems of the theoretical biology): 19–25.

Rosen, Robert; Pattee, Howard; Somorjai, Ray L. 1979. A symposium in theoretical biology. In: Buckley, Paul; Peat, F. David (eds.), *A Question of Physics: Conversations in Physics and Biology*. Toronto: University of Toronto Press, 84–123.

Ruyer, Raymond 1946. *Éléments de psycho-biologie*. Paris: Puf.

Sebeok, Thomas A.; Umiker-Sebeok, Jean (eds.) 1992. *Biosemiotics: Semiotic Web 1991*. Berlin: Mouton de Gruyter.

Stjernfelt, Frederik 1992. Categorical perception as a general prerequisite to the formation of signs? On the biological range of a deep semiotic problem in Hjelmslev's as well as Peirce's semiotics. In: Sebeok, Thomas A.; Umiker-Sebeok, Jean (eds.), *Biosemiotics: Semiotic Web 1991*. Berlin: Mouton de Gruyter, 427–454.

Uexküll, Jakob von 1992 [1934]. A stroll through the worlds of animals and men: A picture book of invisible worlds. *Semiotica* 89(4): 319–391.

Uexküll, Jakob von 1940. *Bedeutungslehre*. (BIOS: Abhandlungen zur theoretischen Biologie und ihrer Geschichte, sowie zur Philosophie der organischen Naturwissenschaften, 10.) Leipzig: Verlag von J. A. Barth.

Wiener, Norbert 1948. *Cybernetics: Or the Control and Communication in the Animal and the Machine*. Cambridge: MIT Press.

Chapter 14
Theories of Signs and Meaning: Views from Copenhagen and Tartu

Jesper Hoffmeyer and Kalevi Kull

Summary. We respond to five questions about the past, present, and future of semiotics. Reflecting our personal experience in biosemiotics, we also try to formulate the role of biological studies for the whole of semiotics.

Peer Bundgaard and Frederik Stjernfelt have asked five questions about the state of the art of semiotic theory from 29 semioticians from around the world.[1] Below are the responses from two of them.

Why were you initially drawn to the theory of signs and meaning?

J.H. As a young professor of biochemistry in the 1970s I felt increasingly appalled by the rhetoric of my fellow experimental biologists. A decade earlier I had become attracted to the discipline of biochemistry because the conception of life as an outcome of dynamic interactions among myriads of truly miniscule particles, cells or molecules, wouldn't stop fascinating me. How could it be the case that 50,000 billion cells — each of which were going to be replaced by new cells in a matter of few weeks — were capable of interacting in such a

---

[1] See the rest in: Bundgaard, Peer; Stjernfelt, Frederik 2009. *Signs and Meaning: 5 Questions*. New York: Automatic Press, VIP.

way as to keep a human being alive day after day through eight decades? Having no inclination for religious explanations I found this to be the most challenging problem for science to solve — far more challenging, in fact, than the much better popularized problems of nuclear physics or space travel. The challenge, by the way, remains unsolved to this day.

But the late 1960s also saw a strong intervention of new politically based kinds of criticisms in university life and, while these influences were obviously far more penetrating in the social sciences, also the natural sciences would gradually be drawn into the battle zone. For science is, of course, deeply involved in technological development and was therefore rightly or wrongly accused of pulling the wires. Thus science was inherently involved in the development of petrochemical agriculture and in the chemical and pharmaceutical industries and could therefore be held responsible for all kinds of problems such as pollution or side effects of medical treatments. Science was also behind new reproductive techniques in the clinical practices, techniques that threatened classical conceptions of individual dignity and integrity, and it was involved in numerous conflicts in the labor market. Finally, scientific conceptions of heritability of intelligence and other personality traits eventually sneaked into the structure of the educational system.

However, none of these accusations challenged science as science, rather they all challenged the way science was used, and scientists understandably rejected most of the criticisms on the grounds that it was unjustified to blame science for the errors committed by way of commercially motivated engineering practices. It nevertheless gradually occurred to me that science was perhaps in some way inherently disposed toward guiding our technical inventiveness in directions that tended to disregard important kinds of human concerns.

The controversy around the so-called "sociobiology" launched by the entomologist Edward O. Wilson in 1975 illustrates this point (Wilson 1975). Basically sociobiology was a new and more sophisticated version of the kind of biological determinism that had already been proposed by students of animal behavior such as Konrad Lorenz and widely popularized by, for instance, Desmond Morris (Lorenz 1974; Morris 1967). Contrary to ethology, however, sociobiology was based on a modern gene-centered understanding of evolutionary theory, and while several competent evolutionary biologists did argue for the inadequacy of the whole approach to support the far-reaching conclusions it claimed (notably Gould and Lewontin 1979), many if not most biologists embraced it as strong support for a genetic determinist conception of human nature. One might well oppose sociobiology for political and religious reasons — it certainly did not leave much space for the contribution of social factors or free will to the configuration of human behavior — but the theory was in many respects nothing but an extension of well-established reasoning patterns in evolutionary biology from the area of morphological or physiological adaptations to the area of animal and human behavior and further on to human mental properties. Opposing sociobiology at the biological level thus required one to search for possible unjustified biases built into the deep structure of the modern gene-based understanding of evolutionary theory.

This is not the place to go into a thorough analysis of this question (for details cf. Hoffmeyer 2008). Suffice it to say that one questionable presupposition that plays a central role in modern biology is the conception of "information flow": according to the so-called "central dogma" the genetic information in a cell is transferred — or "passed on" — from DNA (genes) to RNAs to proteins — never the other way around, i.e., the information flow in the cell is unidirectional. It is, however, not at all clear in what sense this entity, "information", actually flows. Usually genetic information is conceived as a sort of "specification" causing a distinct phenotypic trait, say a missing eye in a salamander or a damaged protein in patients suffering muscular dystrophy. But since the actual biochemistry leading to the phenotypic trait is often quite well known (as, e.g., in the case of eyeless salamanders or muscular dystrophy) it is hard to see what is added to our understanding by introducing the information metaphor. The idea of an "information flow" here simply substitutes for what is usually called causation.

There can be no doubt, however, that the information metaphor does make biological sense. In fact, I don't think that one could understand modern biology at all without the use of "informational language". But whatever this added understanding consists in is not part of cellular reality according to molecular biologists. At the cellular level chemistry exhausts what goes on. The paradox is solved, however, as soon as it is recognized that the heuristic value of the information concept is connected to the role that history (evolution) plays in the life of cells and organisms. What happens is that "history talks", but history is not considered part of biochemistry.

Evolution is an infinitely long series of unique events (e.g., extinctions of species, adaptations, speciations) whereby present organisms take advantage of the methods their ancestors used to get success. By virtue of the genetic "messages" (specifications) carried forward from generation to generation through the reproductive activity of individual organisms, the lineage (i.e., the population seen as an evolving unit) maintains and continuously updates a selective "genomic memory" of its past that in most cases will be a suitable tool for dealing with the future. The lineage in this sense is a historical "subject" capable of interpreting the changing environmental conditions by adjusting the genotype distribution in the next generation. The strangely paradoxical term "natural selection" (paradoxical because nature as such cannot of course "select") may thus be justified to the extent it is understood that the selective agents on this planet are the lineages. As a consequence of this selection process "genetic information" is indeed information in the sense that it is interpreted by the embryos of the present generation as highly structured sets of messages from countless previous generations concerning how to grow and produce functional organisms so that the lineage will survive.

Countering this line of argument one might of course claim an ontology of absolute determinism with the implication that no true interpretative activity actually takes place, but one should be aware that such a claim is not in any way supported by modern science and certainly not by modern biology. Quite to the contrary, since we are here dealing with complex systems obeying so-called chaos dynamics, truly unpredictable contextual influences affect the embryonic interpretation of the hereditary texts in countless ways. And since the interpretation of the genetic information is a contextual process, it gives no meaning to imagine information as an entity that can "flow". Information is nothing like water in a river. In fact, genetic information doesn't move at all, of course; instead, what goes on is that information is interpreted by the organismic systems as holistic subjects.

My dawning understanding through the 1980s of the inadequacy of the biological concept of information thus not only served to legitimize a rejection of sociobiology but, more essentially, it brought me to study the *theory of signs of meaning* — or, in other words, to attempt reframing information biology as *biosemiotics* (Hoffmeyer, Emmeche 1991; Hoffmeyer 1996; 2008).

K.K. The path to semiotics, and particularly to biosemiotics, has had for me at least three guides.

A. Theoretical biology as accompanied with field studies

These were beautiful places — diverse forests, the coast of the Baltic Sea, wooded meadows — where I started the search for a theory of living processes (or theoretical biology) that later turned out to be biosemiotics. Since my high school years, I have spent many summers working as a field biologist, together with scholars who are masters knowledgeable in the diverse life of local ecosystems. The summers from 1968 to 1973 I lived at the Puhtu Biological Station on the east coast of the Baltic Sea, where we studied the ontogenesis of the thermoregulatory behavior in several species of birds. This included ethological studies (both diurnal and nocturnal), laboratory experiments, and telemetrical studies (registration of birds' physiological parameters during flight). Our discussions were focused on a general theory of behavior — it was our intention to develop such a theory. Then I came across Jakob von Uexküll's work. This was followed by modeling work at a forest ecology station, and further, by studies on species coexistence mechanisms, which led to the discovery of the most species-rich plant communities in Europe.

Already during our student years (1970–75) at the University of Tartu, we established regular theoretical biology seminars, and in 1975, the annual Estonian Spring Schools in Theoretical Biology, which continue today (the 35th took place in May 2009). These were influenced by C. H. Waddington's four-part symposia "Towards a Theoretical Biology",[2] which saw general linguistics as a promising paradigm for the theory of general biology. Several of our Spring Schools were focused on the semiotic aspects of biology (for instance, "Theory of Organism (dedicated to Jakob von Uexküll)" (1977), "Theory of Recognition" (1995), "Languages of Life" (1996), "Theory of Communication" (2007), etc.

It appeared that similar groups of young biologists in theoretical biology were established at about the same time in St. Petersburg (led by Sergey Chebanov) and in Moscow (led by Alexander Levich and Alexei Sharov) — all three with an inclination to semiotic approaches. Thus we organized some joint meetings, among these a larger regional conference "Biology and Linguistics" in 1978 in Tartu — which we called biosemiotic. These regional meetings in the late 1970s also meant our study of non-Darwinian biology. This was mainly a structuralist paradigm in biology (of Lev Berg, Alexander Lyubischew, Sergey Meyen, and others),[3] which had been important to grasp as a step towards a truly semiotic biology.

---

[2] See Waddington (1968–1972).

[3] Our views were quite close to the Osaka Group's structuralist biology (see Goodwin *et al.* 1989; Sermonti, Sibatani 1998), there was even one shared member (Lev Beloussov). See also Brauckmann, Kull (1997).

B. Local intellectual traditions

The ecosystem, in which I live, on the East coast of Baltic Sea, is the country where Karl Ernst von Baer, Jakob von Uexküll, and Juri Lotman have lived and worked. This is one of few countries in the world where all intellectuals, at least since the 1970s, know the term *semiotic*. Either by chance, or due to a *genius loci*, or because they belong to the same line in the history of ideas, Baer's, Uexküll's and Lotman's approaches fit each other well, and on their basis, a rich and creative school has formed that has educated the contemporary semioticians in Tartu.

The *Geist* of Baer is in the air in Tartu. This means the strength of an epigenetic view in biology that survived here during its low period (from the 1930s to the 1990s), when genetic preformism dominated in Europe and America. Jakob von Uexküll was a major follower of Baer (see Uexküll 1928; 1940; 1982). My study of his works, and personal contacts with Juri Lotman, ended up with my joining the semiotics department in 1997. In the autumn semester of 1993, I gave my first course in biosemiotics at the University of Tartu, which has turned into a regular course since then. The Jakob von Uexküll Centre was established also in 1993 (Magnus et al. 2006).

The Tartu School of semiotics, as it has developed since the 1960s under the leadership of Juri Lotman, has a particular feature that may probably explain its vitality. This is its interest in the mechanism of the creativity of culture in general (Lotman 1990; 2005; 2009; Torop 2000). The Department of Semiotics at the University of Tartu has grown into an international centre of semiotics that carries on the semiotic traditions of Jakob von Uexküll and Juri Lotman, thus bridging the concepts of umwelt and semiosphere.

C. The biosemiotic circle

The history of biosemiotics can be traced quite far back (Favareau 2007; Kull 1999). However, an international biosemiotic circle has been formed quite recently, in the early 1990s, mainly due to Thomas A. Sebeok and Thure von Uexküll who were able to build a network of people in this field. One can say that it emerged in Glottertal near Freiburg, Germany, where Thure Uexküll and his colleagues organized a workshop on Jakob von Uexküll and biosemiotics, in June 18–20, 1992, and where Jesper Hoffmeyer and myself first met (see Sebeok 2001a; Emmeche *et al.* 2002b). Thomas Sebeok had just edited a volume *Biosemiotics* (Sebeok, Umiker-Sebeok 1992 — the very first book with such a title), and Thure Uexküll had visited biosemiotic workshops both in Tartu (1989) and in Copenhagen (1990). In January 1994, I took a bus to Copenhagen to visit Jesper Hoffmeyer during two weeks, followed by Jesper's visit to Tartu in September. Thus the biosemiotic co-work between the Tartu and Copenhagen groups was established, and this turned out to be a very productive bridge. In 2001, we established the annual international "Gatherings in Biosemiotics" (see Emmeche 2001; Emmeche *et al.* 2002a), and in 2005, the "International Society for Biosemiotic Studies" (see Favareau 2005).

What do you consider your contribution to the field?

J.H. To answer this question very briefly I think my overall contribution to *the theory of signs and meaning* has been to develop biosemiotics as a biologically based general theory of life. The idea of biosemiotics as such was suggested several times throughout the late 20th century (the first time, in fact, as far back as 1962 by the German medical psychologist Friedrich Salomon Rothschild (Rothschild 1962; Kull 1999), and a semiotic approach had been applied to particular areas of biology, zoosemiotics (Sebeok 1963), phytosemiotics (Krampen 1981) or the semiotics of the immune system (Prodi 1988; Sercarz *et al.* 1988). In an important paper from 1984, a group of leading semioticians had agreed on a general statement of a "semiotic perspective on science" as a step towards a new paradigm (Anderson *et al.* 1984), and in 1992 Sebeok and Umiker-Sebeok made biosemiotics the theme of their annual volume in the series *The Semiotic Web* (Sebeok, Umiker-Sebeok 1992). But to the best of my knowledge the paper Claus Emmeche and I published in 1991, "Code-duality and the semiotics of nature" (Hoffmeyer, Emmeche 1991), was the first attempt ever made to make a Peircean semiotic understanding the very basis for the understanding of life as a biological phenomenon, and through the next decades my work has been focused on developing this idea in all its myriad theoretical aspects (Hoffmeyer 1996, 2008).

In a paper from 1995 I tried to formulate the biosemiotic approach to the study of life through nine theses (Hoffmeyer 1995), and building on these theses Claus Emmeche, Kalevi Kull and Frederik Stjernfelt in 2002 summed up my conception of biosemiotics in 13 theses (Emmeche *et al.* 2002b). The first of these states that "Signs, not molecules, are the basic units in the study of life" (Emmeche *et al.* 2002b: 14). The second thesis is that of Code-Duality: "As *analog codes* the organisms recognize and interact with each other in the ecological space, while as *digital codes* ('written' in DNA) they are passively carried forward in time between generations". In this sense, heredity could be seen as "semiotic survival". I went on to suggest that "this principle of code-duality [...] can be taken as a definition of life" (argued in much more detail in Hoffmeyer 2008). Thesis 3 reads: "The simplest entity to possess real semiotic competence is the cell" (Emmeche *et al.* 2002b: 16). In theses 5 and 6 it is claimed that "Subjectness is a more-or-less phenomenon" and that "subjectivity is embodied". Thesis 9 is the thesis of *semethic interaction:* "Wherever a new habit appears, it tends to become a sign for somebody else" (be it a species, a population, an individual organism, or even an organ, gland, or cell). This thesis, in fact, would be one of the backbones in a semiotic understanding of organic evolution. Thesis 11 concerns the idea of a "semiotic niche" to complement the concept of an "ecological niche". And thesis 13 in their summation holds that "biological evolution is a trend toward increased semiotic freedom" (Emmeche *et al.* 2002b: 23).

The biosemiotic conception summarized in these theses presents a strong argument for an emergentist view of life. By semiotic emergence, I mean the establishment of macro-entities or higher-level patterns through a situated exchange of signs between sub-components. The important point here is that while the emergence of higher level patterns may seem to be slightly mysterious (or vitalist) as long as only physical interactions between entities are considered, the same outcome becomes quite understandable when based on *semiotic* interactions among entities at the lower level. Most importantly, *semiotic emergence* in this sense may stand as a possible alternative candidate to natural selection as a mechanism for explaining the evolution of purposive behavior. The prominent example in sociobiological literature of so-called *altruistic behavior* may illustrate this point. The alarm calls of birds or monkeys are said to be altruistic in the sense that the callers themselves are supposed to increase their risk of being caught by a predator. The act of calling lowers the reproductive fitness of the caller and according to the Darwinian scheme any mutant that did not emit these dangerous alarm calls would quickly outcompete the altruists. So, how can altruism (in this sense) nevertheless persist in nature? The answer sociobiologists have given to this question is that evolution is not about the survival of individuals but about the survival of genes. And since every time an altruist gives up its life it will by this very act very probably save the lives of a several of its brothers and sisters (each having a 50% chance of possessing the same altruist gene), the end result is that more copies of the altruist gene will survive than will copies of eventual mutant genes.

This hypothesis may perhaps have some value in the case of primitive animals such as bees or ants. But as soon as we turn to semiotically more sophisticated animals such as birds or mammals it strikes me that the hard problem disappears. "Altruism" in these animals does not necessarily have to be connected to the presence or absence of particular genes but might instead simply be an unavoidable result of an emotional response that follows from the general communicative practices of the concerned animals. Thus, by assuming semiotic interactions among emotional individual organisms to be part of the natural world, many kinds of purposeful behavior patterns might emerge without natural selection being concretely involved in the process. If such emergent behavioral pattern are sufficiently advantageous, natural selection might afterwards be expected to scaffold the patterns by favoring minor genetic adjustments that would facilitate the upholding and transmission of the concerned interaction patterns from generation to generation. But this does not detract from the fact that semiosis, rather than selection, in this case is the key to the evolution of end-directed behavior. And seeing natural semiosis, rather than natural selection, as the motive force behind the evolution of purposive behavior makes a decisive difference. For semiosis inescapably implies an element of Peircean Thirdness, i.e., mediation, whereas natural selection, as presently defined in evolutionary theory, remains safely inside the domain of Secondness. And while the domain of Secondness cannot, logically, evolve to produce creatures with consciousness and first-person experiential worlds, the domain of Thirdness does not preclude, and might perhaps even entail, such an outcome. By assuming semiosic processes to be part of the natural world we might therefore, at least in principle, explain what natural selection is not capable of explaining: The existence in nature of life forms with experiential worlds and even human beings with inherently empathic social needs, moral inclinations and free will (Hoffmeyer 2010).

K.K. Here, a methodological comment is necessary. Describing one's contribution to a field, a modernist view to science — science as anything that is progressing and cumulating — is still often used. The end of modernity in science would mean the acceptance of inquiry as primarily learning, understanding, reformulating, interpreting, and reordering work and activity. This is a shift from a linear and progressing to an ecological and (re)cycling behavior. Semiotics proper is a field of knowledge that characterizes the non-modern, particularly after-modern culture (Deely 2001). This is the culture of (reversible) perfection and learning, not of (irreversible) progress and cumulation. Or, in more biological terms — development, not evolution.

Thus, the question about one's contribution should be reformulated as a question about the key understandings one has arrived at, and about the focuses in one's semiotic work. There are two (central ones) in my case.

A. On the nature of species, or the mechanism of categorization

In biology, the problem of species has survived many generations of biologists. A way towards a semiotic solution has been indicated by the recognition concept of species (Paterson 1993). According to this approach, species is a communicative structure that emerges in communicating populations, assuming that for each individual there exists an individual recognition window that limits the difference between the partners in communication (Kull 1988; 1992). Emerging species is thus a natural category.

Categorization is what all living beings do and what organizes them; categorization is based on recognition processes that are inevitable for organisms; speciation and perceptual categorization are analogical in their mechanism. This is in the nature and origin of species — that species is a self-keeping category *per se*; that species occur because the continuous variability in a communicative population is unstable; that biological species is the same kind of category as any semiosic category (Kull 1992). Thus, what we have here is an explanation of the origin of qualitative differences, or a general mechanism that makes differences, or a process from which the relation of similarity emerges. This leads also to a general definition of semiotics as a study of qualitative diversity.

B. On the reality and realities, or the nature of meaning

A study of the nature of semiosis that includes its inevitable attributes (recognition, memory, feed-forward, code, emergence of absence, etc.) leads to a general model of the life process, a model that explains the emergence of complementarity. A most compact conclusion from this understanding states that *semiosis multiplies reality*, that mind means plurality. Or, synonymously, that life is the local plurality. This means that this "discovery" is also the answer to the question about the nature of life. Life is the phenomenon of the occurrence of plurality in the world. What thus turns out to be locally plural is the reality itself. And this IS life, life itself (Kull 2007).

Since nothing can be simultaneously true, and original or new (i.e., only faults can be original) — this understanding also takes much from predecessors, namely from Jakob von Uexküll's concept of umwelten, from Juri Lotman's concept of non-translatability in translation in semiospheres, from Robert Rosen's analysis of "life itself", and from the Copenhagen concept of complementarity. However, this also uses the concept of semiosis as developed by Charles Peirce, and applies the concept of semiotic threshold via Umberto Eco, Thomas Sebeok, and Juri Lotman (i.e., accepting the existence of both the non-semiotic and the semiotic).

These two points (A and B) appear for me as certain major insights, *heureka*-ideas, or understandings that have led to (or are connected with) many particular "findings" in the interpretation of organic processes and phenomena, and are backed by (or based on) the study of the history and work of other biologists and semioticians.

Besides, there are two other fields I am working in.

C. Semiosic basis of development and evolution

It occurs to me that the principal stages in the development of organic systems are semiosic by their nature. For instance, the development in the sequence of vegetative, animal, and human, can be understood as the jumps over the iconic–indexical and indexical–symbolic threshold (Kull 2009). The biological evolution, approached this way, is non-Darwinian in its basic processes. Organic evolution may even not require much natural selection, because adaptation usually develops prior to its fixation in genetic memory (the latter taking place due to stochastic molecular-genetic processes) (Kull 2000; Hoffmeyer, Kull 2003).

D. Conceptualization of biosemiotics

I have been interested in the history of biosemiotics. No, this is not an explanation via history — the reason to study history should not be based on a historical explanation. Studying history is the same as studying a life process. Life's explanation should not lie in the past, it rather relies on something that reminds us of the future, i.e., in void or absence.

Thus I have tried to rethink the history of biology as something that is broader than its modernist account. This includes an analysis of the three major archetypes of biological thinking (ladder, tree, web) (Kull 2003); an account of Jakob von Uexküll's work (Kull 2001); the placement of particular works in the history of biosemiotic ideas (Kull 1999).

What is the proper role of a theory of signs and meaning in relation to other academic disciplines?

J.H. From what has already been said it is obvious that I consider the theory of signs and meaning to be fairly fundamental. It in no way substitutes for the individual life sciences (from biochemistry to medicine) but it offers an indispensable resource base for the necessary work of reframing theoretical structures of the life sciences to accommodate the semiotic character of their subject matter. In this process, however, the theory must itself accommodate to the changed scope of its approach (more on this below).

In a broader perspective the theory of signs of meaning should serve as a realistic input to the social and human sciences in general. As T. L. Short has recently formulated it, discussing Peirce's theory of signs, the philosophical significance of the theory is "to construct a naturalistic but non-reductive account of the human mind, and to explain and defend the claim that the sciences are objective in their mode of inquiry and in fact yield knowledge of an independently existing reality" (Short 2007: ix). By offering a realistic input to the humanities, the extended theory of signs and meaning may once again serve to turn the focus of these sciences towards the implications it has for all kinds of human significative activity that such activity springs from bodies of "flesh and blood" — or, more precisely, from an incredibly intricate interaction between different tissues and cells in the semiotically integrated psycho-neuro-immuno-endocrine landscape of the body. Such an endeavor may well run somewhat counter to the hegemonial trend in these sciences for the last three of four decades. For, again citing Short: "Continental writers, approaching Peirce from a background of Saussurean semiology, have systematically misrepresented his semeiotic. For the two doctrines are fundamentally incompatible [...] The unholy union of Saussure's supposed conventionalism with the breadth of Peirce's mature semeiotic gave bastard birth to an extreme relativism and irrealism — a modern version of sophistry that Saussure and Peirce would both have rejected" (Short 2007: xiii).

K.K.

A. Semiotics is complementary

A distinction made by John Locke is remarkable — the three main fields of knowledge are physics, semiotics, and ethics. This is, in a way, even beyond Peirce, because Peirce was not much interested in physics, or, more precisely, Peirce did not believe in the universality of physical laws, and thus deduced all methodology from life sciences, or logic. Thus the role and relatedness to other disciplines is still a problem.

Semiotics and physics can be seen as the two basic complementary methodologies, the first for qualitative, and the second for quantitative inquiry (Kull 2007). Semiotics deals with the pluralist and incommensurable, whereas physics is the approach for the convertible and monistic.

Pointing to this most meaningful difference, it is necessary to admit that it is certainly possible to understand and to explain the moves from one to another. That means it is possible to explain the emergence of semiosis from non-semiosis (as life from non-life) on the basis of physical description,[4] and reciprocally, the collapse from semiotic to physical on the basis of semiotic language.[5]

B. Semiotics is fundamental

If the major advances in the understanding of the world (as both developmental and evolutionary biosemiotics aim to describe), and the principal steps in the history of philosophy (as John Deely has attempted to demonstrate, in Deely 2001) are directly related to the advancements in semiosis and its (self)understanding, then Sebeok's global semiotic approach (Sebeok 2001b) turns out to be fundamental indeed. Semiotics is fundamental also in the sense that it studies (and can explain) the principal processes of life, mind, and culture (Anderson et al. 1984).

C. Semiotics is practically applicable

Semiotics is practically applicable in order to study and analyze particular cases. Any particular phenomenon of life or culture can serve as an object for a semiotic inquiry. Accordingly, there (should) exist practical methods of semiotic analysis. Different semiotic schools have already developed some versions of these methods.

What do you consider the most important topics and/or contributions in the theory of meaning and signs?

J.H. Presently the semiotics of natural processes occupies a very marginal position in the general conception of semiotics — whether seen from outside or from inside the field. This is partly to be blamed

---

[4]See a list, for instance, in Yates (1992).
[5]A possible example, see Kauffman (2005).

on tradition, but it also reflects the simple fact that the sophistication of human semiotic activity is radically more pronounced than that of any other animal and, besides, that human semiotic activity for good reasons concerns us a lot more than does that of other species.

Not surprisingly, I suppose, I will nevertheless recommend an increased attention to the consequences it has for semiotics that signs and meanings are rooted in bodies and, in the end, evolutionary dynamics. Nothing is subtracted from human semiotic freedom by the recognition that this freedom has earthly grounding. The fact that we cannot fly has been, and will in all future be, dealt with in human imagination and expressions in countless and truly unpredictable ways, but it remains a fact that our missing ability to fly does make up for a universal constraint that every human person must somehow accommodate to mentally. Biologists are in no particular position to specify what kind of more or less hidden universals in human nature that play games with our lives, for the essence of the fitness advantage it gave our remote ancestors to evolve linguistically competent minds, exactly was to get rid of those hereditary restrictions on behavior that biologists are schooled to know about. Or, as Susanne Langer saw so many years ago:

> There is a primary need in man, which other creatures probably do not have, and which actuates all his apparently unzoological aims, his wistful fancies, his consciousness of value, his utterly impractical enthusiasms, and his awareness of a "Beyond" filled with holiness [...] *this basic need, which certainly is obvious only in man, is the need of symbolization.* The symbol-making function is one of man's primary activities, like eating, looking, or moving about [...] it is not the essential act of thought that is symbolization, but an act *essential to thought*, and prior to it. Symbolization is the essential act of mind; and mind takes in more than what is commonly called thought. (Langer 1942: 32–33)

But the fact that our semiotic life has evolutionary roots does of course influence the way we communicate and understand each other and this ought to be conceptualized by semiotics. For instance, it immediately explains the apparent universal that we all feel deeply attracted to other semiotically free animals such as dolphins or apes whereas we rarely take pity on reptiles or fishes. The very fact that the psychological and corporeal reality of each individual is so tightly interconnected implies that the aesthetic and ethical needs of a human being can not any longer be coherently understood as "incorporeal" — which is to say that such needs cannot any longer be conceived as being based only on "intellectual willpower" but rather as always arising from a conjoined corporeal/intellectual effort. Accordingly, the role played by "intellectual judgment" in this context is that of a *guide* rather than that of an *executive officer*. This does not, of course, imply a reduction of ethics to brute ethology — for the whole point of this argument is that the semiotization of the body has already pushed the potential for genuinely free action back *into* corporeal nature. Human semiotic life is enabled by its corporeal basis just as much as our corporeal life is in a deep sense semiotic. The corporeality of the human being is therefore, for better or worse (for, not only the human good, but also the human evil, has human empathy as its precondition), the key to our ability to empathize with "the Other". This is not because the other *has* a body, but because he or she *is* a body, just like I *am* a body myself.

By emphasizing the naturalistic aspect of the theory of signs and meaning I am not attempting to make any judgments as to which aspect of such a theory is the more important one. My concern rather is for semiotics to consider the numerous implications of this anchoring of the theory in the realities of "flesh and blood". Not only is there no such thing as a free lunch, there isn't anything like a free fantasy either. Even fantasy is embedded in bodily frames — which, however, does not make our fantasies one bit less entertaining.

K.K. What occurs to me as the fundamental achievements and results of semiotics, include the following.

A. The basic models of semiosis

There are several of these, particularly the triadic model by Charles S. Peirce, the dyadic one by Ferdinand de Saussure and his school, the cyclic (feed-forward) model by Jakob von Uexküll, the communicational one by Roman Jakobson, and the semiospheric conception by Juri Lotman. These are complementary models, each having its own separate value and applicability.

B. The concept of semiotic thresholds

As a concept of lower semiotic threshold it has been introduced by Umberto Eco (1976), then developed and (re)placed by Thomas Sebeok and other biosemioticians (Anderson *et al.* 1984; Sebeok 2001b; Emmeche *et al.* 2002b), and further applied by Terrence Deacon (as the symbolic threshold, Deacon 1997). The lower semiotic threshold distinguishes between semiosic and non-semiosic, whereas the indexical and symbolic threshold may allow us to distinguish between the vegetative, animal, and human (propositional) semiosis.

C. The development of biosemiotics, zoosemiotics, and semiotics of culture

Both as clearly separated fields, and as parts of the same semiotics (Hoffmeyer 1996; 2008). Very much due to these fields, it has been possible to overcome the superficial divides between mind and body, and culture and nature, and to view both organism and culture as semiosic processes.

D. Turning semiotics into semiotics

This has been very much Thomas A. Sebeok's enterprise and his major achievement. This is why we have this field recognized as one (and as many).

What are the most important open problems in this field and what are the prospects for progress?

J.H. One important question that divides people in semiotics is the question often referred to as the "semiotic threshold", i.e., the problem of defining the simplest system capable of semiosic activity. Personally I have suggested the living cell as the simplest existing semiotic entity on the grounds that the cell is the simplest system possessing the twin properties of self-referential activity (based on DNA) and other-referential activity (based on surface receptors coupled to mechanisms for intra-cellular signal-transduction) (Hoffmeyer 1995). This essentially agrees with Sebeok's claim for semiosis and life to be co-existent (Sebeok 1985). Thomas L. Short opposes this conception since semiosis or interpretation presupposes an ability for purposeful action and thus for making mistakes; a capacity that, in his view, limits interpretative activity to animals (Short 2007). An evolutionary counterargument to Short's view might address the question of scales and levels here. Since evolution, as I hold, has an inherent trend toward the production of systems with an ever increasing capacity for semiotic freedom, one might suggest that the "semiotic subject", i.e., the living system responsible for doing the interpretations (and thus possibly misinterpretations) has not always been the individual organism. At earlier stages of the evolutionary process semiotic agency should more accurately be ascribed to the lineage, i.e., the evolving species. While individual organisms at this stage of evolution might not yet have acquired the full capacity for purposeful intentional activity, the lineage as such does indeed exhibit a capacity for doing mistakes (followed often by extinction in this case). Only in later stages of evolution, when bigger brains had evolved, was this capacity taken over by individual organisms capable of coping with the world in much more fallible ways than e.g., individual insects or plants.

These are matters for future clarification, as are the merits of a more radical position such as John Deely's claim for a general physiosemiosis (Deely 1990; Salthe 1999). But it seems more and more urgent for biosemiotics to address the question of thresholds in a more nuanced way. For all we know, organic evolution presupposes some sort of continuity — simply because organisms in one generation necessarily must be very much like the organisms in the preceding generation — all the way back to the first organisms in the world. So-called hopeful monsters — saltatory creations of radically different organisms/species — do not seem to occur, and neither has any credible mechanism for such events been suggested. And yet, entirely new properties have certainly emerged in the course of evolution. To mention just a few of the more pronounced cases: photosynthesis, multicellularity, sexual reproduction, appearance of immune systems, nervous systems and, of course, of consciousness. In all such cases the threshold problem poses itself, and yet it might be more fruitful to circumvent this recurring problem and instead search for or define the graded series of emergent events that supposedly have led from the most simple forms of sentience or irritability in living systems to more and more advanced expressions of life's agency or "striving" (to use Darwin's own term) such as, e.g., intentionality, possession of emotional tones, of experiential world, or of consciousness.

K.K. Let me point out just two that are, however, extensive.

A. The biosemiotic project

This is indeed a pretty ambitious enterprise — to put biology on a semiotic basis. It will include rethinking much of biological theory, and a reworking of biological methodology. It means a reformulation and interpretation of biological knowledge on the basis of semiosis. And this is a full-scale introduction of qualitative methods into biological research. Particularly, this will include a deep inquiry into the modeling and analysis of the main attributes of life — activity, needs, intentions, memory, categorization, etc. As a result, it will also give a deeper understanding and description of semiotic thresholds, including the threshold that is responsible for being a human.

B. The problem of semiotic balance

Stability/instability of semiosic processes includes a peculiar set of phenomena, and its importance stems from the importance of preserving diversity and humanity. The study of organic balance will be a guide and the basis here. This is because the balance of life (which is an assumption for an ecological balance) by its very nature is a semiosic balance. Which means that the problem of peace, of the balance of cultures, will converge with the problems of ecological balance and the problem of health; thus the protection of biodiversity and protection of cultural diversity turn out to be the same — the protection of diversity, or quality as such.

References

Anderson, Myrdene; Deely, John; Krampen, Martin; Ransdell, Joseph; Sebeok, Thomas A.; Uexküll, Thure von 1984. A semiotic perspective on the sciences: Steps toward a new paradigm. *Semiotica* 52(1/2): 7–47.
Baer, Karl Ernst von 1864. *Reden gehalten in wissenschaftlichen Versammlungen und kleinere Aufsätze vermischten Inhalts*. Erster Theil. St. Petersburg: H. Schmitzdorff.
Brauckmann, Sabine; Kull, Kalevi 1997. Nomogenetic biology and its western counterparts. In: Naumov, Rem V.; Marasov, Anatoli N.; Gurkin, Vladimir A. (eds.), *Lyubischevskie Chteniya 1997*. Ul'yanovsk: Ul'yanovskij gosudarstvennyj pedagogicheskij universitet, 72–77.
Deacon, Terrence 1997. *The Symbolic Species*. London: Penguin.
Deely, John 1990. *Basics of Semiotics*. Bloomington: Indiana University Press.
Deely, John 2001. *Four Ages of Understanding*. Toronto: University of Toronto Press.
Eco, Umberto 1976. *A Theory of Semiotics*. Bloomington: Indiana University Press.
Emmeche, Claus 2001. The emergence of signs of living feelings: Reverberations from the first Gatherings in Biosemiotics. *Sign Systems Studies* 29(1): 369–376.
Emmeche, Claus; Hoffmeyer, Jesper; Kull, Kalevi 2002a. Editors' comment. *Sign Systems Studies* 30(1): 11–13.
Emmeche, Claus; Kull, Kalevi; Stjernfelt, Frederik 2002b. *Reading Hoffmeyer, Rethinking Biology*. [*Tartu Semiotics Library* 3.] Tartu: Tartu University Press.
Favareau, Donald 2005. Founding a world biosemiotics institution: The International Society for Biosemiotic Studies. *Sign Systems Studies* 33(2): 481–485.
Favareau, Donald 2007. The evolutionary history of biosemiotics. In: Barbieri, M. (ed.), *Introduction to Biosemiotics*. Berlin: Springer, 1–67.

Goodwin, Brian; Sibatani, Atuhiro; Webster, Gerry (eds.), 1989. *Dynamic Structures in Biology*. Edinburgh: Edinburgh University Press.
Gould, Stephen Jay; Lewontin, Richard C. 1979. The spandrels of San Marco and the Panglossian Paradigm: A critique of the adaptationist programme. *Proceedings of the Royal Society of London* B 205(1161): 581–598.
Hoffmeyer, Jesper 1995. The swarming cyberspace of the body. *Cybernetics and Human Knowing* 3: 16–25.
Hoffmeyer, Jesper 1996. *Signs of Meaning in the Universe*. Bloomington: Indiana University Press.
Hoffmeyer, Jesper 2008. *Biosemiotics: An Examination into the Signs of Life and the Life of Signs*. Scranton: Scranton University Press.
Hoffmeyer, Jesper 2010. A biosemiotic approach to the question of meaning. *Zygon* 45: 367–390.
Hoffmeyer, Jesper; Emmeche, Claus 1991. Code-duality and the semiotics of nature. In: Anderson, Myrdene; Merrell, Floyd (eds.), *On Semiotic Modeling*. New York: Mouton de Gruyter, 117–166.
Hoffmeyer, Jesper; Kull, Kalevi 2003. Baldwin and biosemiotics: What intelligence is for. In: Weber, Bruce H.; Depew, David J. (eds.), *Evolution and Learning: The Baldwin Effect Reconsidered*. Cambridge: MIT Press, 253–272.
Kauffman, Louis H. 2005. Virtual logic — the one and the many. *Cybernetics and Human Knowing* 12(1/2): 159–167.
Krampen, Martin 1981. Phytosemiotics. *Semiotica* 36: 187–209.
Kull, Kalevi 1988. The origin of species: A new view. In: Kull, Kalevi; Tiivel, Toomas (eds.), *Lectures in Theoretical Biology*. Tallinn: Valgus, 73–77.
Kull, Kalevi 1992. Evolution and semiotics. In: Sebeok, Thomas A.; Umiker-Sebeok, Jean (eds.), *Biosemiotics: Semiotic Web 1991*. Berlin: Mouton de Gruyter, 221–233.
Kull, Kalevi 1999. Biosemiotics in the twentieth century: A view from biology. *Semiotica* 127(1/4), 385–414.
Kull, Kalevi 2000. Organisms can be proud to have been their own designers. *Cybernetics and Human Knowing* 7(1): 45–55.
Kull, Kalevi 2001. Jakob von Uexküll: An introduction. *Semiotica* 134(1/4): 1–59.
Kull, Kalevi 2003. Ladder, tree, web: The ages of biological understanding. *Sign Systems Studies* 31(2): 589–603.
Kull, Kalevi 2007. Biosemiotics and biophysics — the fundamental approaches to the study of life. In: Barbieri, Marcello (ed.), *Introduction to Biosemiotics: The New Biological Synthesis*. Berlin: Springer, 167–177.
Kull, Kalevi 2009. Vegetative, animal, and cultural semiosis: The semiotic threshold zones. *Cognitive Semiotics* 4: 8–27.
Langer, Susanne 1942. *Philosophy in a New Key*. Cambridge: Harvard University Press.
Lorenz, Konrad 1974. *On Agression*. New York: Harcourt Brace Jovanovich.
Lotman, Juri M. 1990. *Universe of the Mind: A Semiotic Theory of Culture*. London: I.B.Tauris.

Lotman, Juri M. 2005 [1984]. On the semiosphere. *Sign Systems Studies* 33(1): 215–239.
Lotman, Juri M. 2009. *Culture and Explosion*. Berlin: Mouton de Gruyter.
Magnus, Riin; Maran, Timo; Kull, Kalevi 2006. Jakob von Uexküll Centre, since 1993. *Sign Systems Studies* 34(1): 67–81.
Morris, Desmond 1967. *The Naked Ape*. London: Jonathan Cape.
Paterson, Hugh E.H. 1993. *Evolution and the Recognition Concept of Species*. Baltimore: Johns Hopkins University Press.
Prodi, Giorgio 1988. Signs and codes in immunology. In: Sercarz *et al*. 1988: 53–64.
Rothschild, Friedrich Salomon 1962. Laws of symbolic mediation in the dynamics of self and personality. *Annals of New York Academy of Sciences* 96: 774–784.
Salthe, Stanley 1999. A semiotic attempt to corral creativity via generativity. *Semiotica* 127(1/4): 481–495.
Sebeok, Thomas A. 1963. Communication in animals and men. *Language* 39: 448–466.
Sebeok, Thomas A. 2001a. Biosemiotics: Its roots, proliferation, and prospects. *Semiotica* 134(1/4): 61–78.
Sebeok, Thomas A. 2001b. *Global Semiotics*. Bloomington: Indiana University Press.
Sebeok, Thomas A.; Umiker-Sebeok, Jean (eds.) 1992. *Biosemiotics: The Semiotic Web 1991*. Berlin: Mouton de Gruyter.
Sercarz, Eli E.; Celada, Franco; Mitchison, N. Avrion; Tada, Tomio (eds.) 1988. *The Semiotics of Cellular Communication in the Immune System*. Berlin: Springer.
Sermonti, Giuseppe; Sibatani, Atuhiro 1998. The 'Osaka Group': Ten years later. *Rivista di Biologia* 91: 125–158.
Short, Thomas L. 2007. *Peirce's Theory of Signs*. Cambridge: Cambridge University Press.
Torop, Peeter 2000. New Tartu semiotics. *European Journal for Semiotic Studies* 12(1): 5–22.
Uexküll, Jakob von 1928 [1920]. *Theoretische Biologie*. 2te Aufl. Berlin: J. Springer.
Uexküll, Jakob von 1940. *Bedeutungslehre*. Leipzig: J.A. Barth.
Uexküll, Jakob von 1982 [1940]. The theory of meaning. *Semiotica* 42(1): 25–82.
Waddington, Conrad Hal 1968–1972. *Towards a Theoretical Biology*. Vols. 1–4. Edinburgh: Edinburgh University Press.
Wilson, Edward O. 1975. *Sociobiology: The New Synthesis*. Cambridge: Belknap Press.
Yates, F. Eugene 1992. On the emergence of chemical languages. In: Sebeok, Thomas A.; Umiker-Sebeok, Jean; Young, Evan P. (eds.), *Biosemiotics: The Semiotic Web 1991*. Berlin: Mouton de Gruyter, 471–486.

Acknowledgements

We are very thankful to all our friends and colleagues. We have prepared this book for your delight.

Some of the earlier versions of the texts here have been published in the following sources. (We appreciate the copyright permissions from all colleagues and publishers. *We also thank the European Regional Development Fund, Center of Excellence CECT, SF0182748s06, ETF7790, 8403.*)

Emmeche, Claus 2004. A-life, organism and body: The semiotics of emergent levels. In: Bedau, Mark; Husbands, Phil; Hutton, Tim; Kumar, Sanjev; Suzuki, Hideaki (eds.), *Workshop and Tutorial Proceedings. Ninth International Conference on the Simulation and Synthesis of Living Systems (Alife IX), Boston Massachusetts, September 12th, 2004.* Boston, 117–124.

Emmeche, Claus; Hoffmeyer, Jesper; Kull, Kalevi; Markoš, Anton; Stjernfelt, Frederik; Favareau, Donald 2008. The IASS roundtable on biosemiotics: A discussion with some founders of the field. *The American Journal of Semiotics* 24(1/3): 1–21.

Hoffmeyer, Jesper 2009. Biology is immature biosemiotics. In: Deely, John; Sbrocchi, Leonard G. (eds.), *Semiotics 2008: Specialization, Semiosis, Semiotics.* (Proceedings of the 33rd Annual Meeting of the Semiotic Society of America, Houston, TX, 16–19 October 2008.) Ottawa: Legas, 927–942.

Hoffmeyer, Jesper 2009. [Interview.] In: Bundgaard, Peer; Stjernfelt, Frederik (eds.), *Signs and Meaning: 5 Questions.* New York: Automatic Press, VIP, 61–70.

Kotov, Kaie; Kull, Kalevi 2006. Semiosphere versus biosphere. In: Brown, Keith (ed.), *Encyclopedia of Language and Linguistics*, vol. 11. Oxford: Elsevier, 194–198.

Kull, Kalevi 2007. Life is many: On the methods of biosemiotics. In: Witzany, Günther (ed.), *Biosemiotics in Transdisciplinary Contexts: Proceedings of the Gathering in Biosemiotics 6, Salzburg 2006.* Salzburg: Umweb, 193–202.

Kull, Kalevi 2009. Semiosic means alive: A Tartu view. In: Bundgaard, Peer; Stjernfelt, Frederik (eds.), *Signs and Meaning: 5 Questions.* New York: Automatic Press, VIP, 113–120.

Acknowledgements

Kull, Kalevi; Emmeche, Claus; Favareau, Donald 2008. Biosemiotic questions. *Biosemiotics* 1(1): 41–55.

Kull, Kalevi; Deacon, Terrence; Emmeche, Claus; Hoffmeyer, Jesper; Stjernfelt, Frederik 2009. Theses on biosemiotics: Prolegomena to a theoretical biology. *Biological Theory* 4(2): 167–173.

Maran, Timo 2007. Structural and semiotic aspects of biological mimicry. In: Witzany, Günther (ed.), *Biosemiotics in Transdisciplinary Contexts: Proceedings of the Gathering in Biosemiotics 6, Salzburg 2006*. Salzburg: Umweb, 237–243.

Neuman, Yair 2005. Why do we need signs in biological systems? *Rivista di Biologia: Biology Forum* 98: 497–512.

Pattee, Howard H.; Kull, Kalevi 2009. A biosemiotic conversation: Between physics and semiotics. *Sign Systems Studies* 37(1/2): 311–331.

Turovski, Aleksei 2000. The semiotics of animal freedom: A zoologist's attempt to perceive the semiotic aim of H. Hediger. *Sign Systems Studies* 28: 380–387.

Name Index

Adami, Chris 109
Alberts, Bruce 197, 208
Alexandrov, Vladimir E. 184, 191
Alford, Ross A. 163, 164
Allen, Colin 119, 127
Alterman, Richard 111
Andersen, Peter B. 69, 85
Anderson, Myrdene 25, 39, 69, 81, 85, 125, 127, 192, 208, 261, 270, 278, 281
Andrade Pérez, Luis E. 14, 16
Andrews, Edna 191
Arnellos, Argyris 86
Augustyn, Prisca 13, 17
Ayres, Peter G. 151, 159, 165
Azcon, Rosario 164
Azcon-Agilar, Concepción 164

Bacon, Francis 204
Baer, Karl E. von 9, 227, 269, 284
Bakhtin, Mikhail M. 193, 208
Baldwin, Ian T. 147, 151, 153, 158, 164
Baldwin, James M. 87, 140, 261, 285
Baluška, Frantisek 83, 85
Barbieri, Marcello 11, 13, 14, 17, 18, 40, 82, 85–89, 103, 104, 109, 115, 127, 128, 164, 223, 232, 284, 285
Barea, José-Miguel 150, 164
Barlow, Peter W. 9, 17, 83, 86
Bartel, David P. 111
Barthes, Roland 3, 12, 17
Bastin, Ted 232
Bates, Henry W. 169, 176
Bateson, Gregory 19, 81, 86, 87, 115, 128, 163, 164, 180, 191, 202, 206, 208, 250

Bateson, William 8
Beckage, Nancy E. 141
Becker, Anton L. 207, 208
Bedau, Mark 86, 92, 109, 111, 127, 287
Bekoff, Marc 119, 127
Bellgard, Matthew I. 160, 164
Beloussov, Lev 268
Bennett, Charles H. 199, 204, 208
Bense, Max 14, 17
Berg, Lev 9, 227, 268
Bohr, Niels 121, 226
Borges, Jorge Luis 203, 204, 208
Borup, Mia Trolle 109
Bouissac, Paul 5, 17, 90
Bower, Erica 151, 162, 165
Brauckmann, Sabine 268, 284
Bray, Dennis 208
Brier, Soren 14, 17
Brower, Jane Van Zandt 169, 176
Brower, Lincoln P. 169, 176
Brown, Keith 287
Brown, Valerie K. 164–166
Bruin, Jan 153–155, 157, 164, 165
Bruni, Luis E. xi, 81, 86, 143, 145, 148, 149, 153, 154, 156–158, 164
Buckley, Paul 262
Bundgaard, Peer 263, 287
Bunge, Gustav 71, 86
Buss, Leo W. 103, 109
Bycroft, Barrie W. 165, 166

Cangelosi, Angelo 15, 17
Caporale, Lynn H. 164
Carlier, Pascal 168, 177
Carr, Gerald F. 192, 193

Cassirer, Ernst 9, 11, 17, 19
Celada, Franco 18, 90, 286
Chaitin, Gregory 260
Chalmers, David J. 77, 86
Chandler, Jerry L. R. 110
Chang, Han-Liang 184, 191
Chang, Jeff H. 166
Chebanov, Sergey V. 12, 17, 115, 119, 125, 127, 227, 240, 268
Chen, Liaohai 111
Chesterton, Gilbert K. 133, 136, 141
Chhabra, Siri Ram 165
Christiansen, Gunna 165
Christiansen, Peder V. 85
Chuang, Isaac L. 204, 209
Cimatti, Felice 14, 17
Clardy, Jon 197, 208
Clark, Andy 77, 86
Clarke, David S. 79, 86, 123, 127
Clay, Keith 159, 160, 165
Clayton, Philip 40, 55, 65
Cohen, Irun R. 196, 208
Cohen, Irun R. 196, 208
Cooper, William S. 13, 17
Cott, Hugh B. 169, 177
Crick, Francis 197, 237
Curie, Pierre 186
Cuvier, Georges 8

Danesi, Marcel 170, 177
Dangl, Jeffery L. 166
Darwin, Charles 8, 16, 50–53, 65, 254
Davies, Nicholas B. 88
Davies, Paul 40, 65
Dawkins, Richard 74, 88, 207, 208
Deacon, Terrence xi, 14, 17, 25–27, 29, 35, 40, 55, 65, 75, 79, 82, 83, 85, 86, 281, 284, 288
Deamer, David 111
Dean, C. Garfield 128
Deely, John 10, 15, 17, 19, 39, 44, 52, 65, 73, 75, 83, 86, 116, 127–129, 189, 191, 223, 261, 273, 278, 282–284, 287
Défago, Geneviève 150, 165
Del Prete, Sandro 117
Delbrück, Max 226
Deleuze, Gilles 200, 208

Delpos, Manuela 111, 192
Dennett, Daniel 7
Depew, David J. 7, 17, 87, 261, 285
Dewsbury, Donald A. 141
Dicke, Marcel 153–155, 157, 164, 165
Dilthey, Wilhelm 13
Donk, Ellen van 166
Dosse, François 11, 17
Driesch, Hans A. E. 8, 9
Dunny, Gary M. 166
Dwyer, Peter D. 7, 18

Eberl, Leo 150, 165
Eco, Umberto 3, 18, 79, 86, 275, 281, 284
Eddington, Arthur S. 217, 232
Edwards, Simon G. 166
El-Hani, Charbel N. 14, 18, 81, 82, 86, 93, 95, 98, 102, 103, 109, 168, 177
Elsasser, Walter 119, 127, 214, 232
Emmeche, Claus x, xi, 1, 2, 14, 18, 25, 26, 40, 65, 67, 69, 73, 76, 79, 82, 85–87, 89, 91–93, 96, 97, 103, 109, 110, 115, 127, 154, 165, 181, 192, 206, 208, 228, 229, 232, 235–239, 242, 245, 250, 258–261, 267, 270, 271, 281, 284, 285, 287, 288
Epel, David 255, 261

Faeth, Stanley H. 151, 165
Falkow, Stanley 163, 165
Faria, Marcella 82, 87
Farina, Almo 81, 87
Favareau, Donald xi, 2, 15, 18, 25, 40, 67, 73, 82, 83, 87, 116, 127, 193, 229, 232, 235, 236, 238, 239, 241–252, 255–257, 259, 261, 269, 270, 284, 287, 288
Fellowes, Mark D. E. 177
Finnemann, Niels O. 85
Florkin, Marcel 11, 18, 154, 227, 232
Fokkema, Nyckle J. 159, 165
Fontana, Walter 13, 18
Forbus, Ken D. 125, 127

Gadamer, Hans-Georg 13, 125
Galitski, Timothy 87
Gange, Allan C. 151, 162, 164, 165, 166

Gilbert, Francis 175, 177
Gilbert, Scott F. 84, 87, 93, 110, 255, 261
Givskov, Michael 165, 166
Goodenough, Ward H. 6, 18
Goodwin, Brian 7, 9, 18, 19, 21, 268, 285
Gould, Stephen J. 8, 18, 45, 65, 265, 285
Grant, Bruce 140, 141
Greco, Alberto 17
Green, David G. 109
Griesemer, James R. 97, 111
Grygar, Filip 261
Gurkin, Vladimir A. 284
Gutmann, Mathias 10, 18.
Günther, Klaus 145

Haber, Wolfgang 144–146, 165
Hacking, Ian 67, 87
Hajnal, László 261
Halitschke, Rayko 164
Haraway, Donna 105, 110
Harnad, Stevan 15, 17, 18
Harries-Jones, Peter 81
Hartsoeker, Nicolaas 54
Harvey, Jeffrey A. 166
Hatcher, Paul E. 151, 159, 165, 166
Hawkes, Terence 11, 12, 18
Hediger, Heini 133–136, 138–141, 288
Heidegger, Martin 125
Heip, Carlo H. R. 166
Herzfeld, Michael 20, 90
Hertz, Heinrich 228, 229, 232
Hippocrates 219
Hoffmann, E. T. A. 139, 141
Hoffmeyer, Jesper x, xi, 1, 4, 14, 15, 18,
   19, 25, 26, 34, 40, 41, 43, 48, 49, 51, 59,
   63, 65, 69, 73, 75, 78, 81, 85–87, 94,
   104, 109–111, 115, 127, 128, 145, 154,
   164, 165, 179, 181, 189–192, 196, 206,
   208, 216, 228, 232, 235, 236, 238–240,
   242–244, 247–249, 251, 253, 256, 260,
   261, 263–288
Hofstadter, Douglas R. 105, 110
Holloway, Graham J. 177
Holquist, Michael 195, 196, 208
Honegger, René E. 134, 141
Hood, Leroy 87
Howie, James W. 161, 165

Huber, Johannes 82, 87
Hulswit, Menno 95, 110
Husbands, Phil 86, 127, 287
Hutton, Tim 86, 127, 287

Ideker, Trey 84, 87
Ikegami, Takashi 109
Imanishi, Tadashi 164
Itoh, Takeshi 164
Ivanov, Vyacheslav V. 122, 128, 181,
   189, 192

Jablonka, Eva 221, 232
Jakobson, Roman 5, 9, 19, 198, 208,
Jiménez-Montano, Miguel A. 9, 19
Johnson, Alexander 208
Johnson, Dennis 208
Johnson, Mark 75, 88
Jones, Jonathan D. G. 155, 165

Kaneko, Kunihiko 109
Kant, Immanuel 31, 40
Karban, Richard 151, 165
Kauffman, Louis H. 122, 128, 278, 285
Kauffman, Stuart 45, 55, 64, 65, 102,
   110, 193, 260
Kaufmann, Morgan 127
Kaushik, Meena 125, 128
Kawade, Yoshimi 184, 192
Keel, Cristoph 150, 165
Keller, Evelyn F. 67, 87
Kember, Sarah 105, 110
Keskpaik, Riste 80, 87
Kessler, Andre 164
Kinji, Imanishi 192
Kirsh, David 111
Kjelleberg, Staffan 166
Kleisner, Karel 168, 170, 177, 261
Kohler, Robert E. 107, 110
Køppe, Simo 109
Kotov, Kaie xii, 179, 184, 192, 287
Kowzan, Tadeusz 170, 177
Krakauer, David C. 111
Krampen, Martin 2, 19, 39, 73, 81–83,
   85, 87, 88, 261, 270, 284, 285
Kratochwil, Anselm 145, 161, 162, 165
Kratochvil, Zdenek 261

Krebs, John R. 74, 88
Krois, John Michael 11, 19
Kruse, Felicia 129
Kull, Kalevi x, xii, 1–3, 6, 13, 15, 18, 19, 25, 26, 33, 40, 55, 65, 67, 69, 72, 73, 75, 78, 79, 83, 87, 88, 94, 95, 98, 110, 113, 115, 120, 122, 124, 126–128, 140, 141, 179, 182, 184, 190, 192, 196, 209, 213–232, 235–256, 261, 263–288
Kumar, Sanjev 86, 127, 287

Lakoff, George 75, 88
Lamb, Marion J. 221, 232
Lambert, David M. 76, 88, 140, 141
Landauer, Rolf 199, 204, 208, 209
Langer, Susanne K. 279, 285
Langton, Cristopher G. 91, 110, 233
Laplace, Pierre-Simon 217
Latour, Bruno 105, 108, 110
Laughlin, Robert 64, 65
Lestel, Dominique 80, 88
Lester, Mark 208
Levich, Alexander 268
Levins, Richard 67, 88
Lévi-Strauss, Claude 8, 11, 19
Levit, George S. 185, 187, 192
Lewis, Julian 208
Lewontin, Richard C. 67, 88, 265, 285
Liszka, James J. 95, 110
Locke, John 277
Lorenz, Konrad 134, 265, 285
Lotman, Juri 118, 128, 179–184, 189–193, 269, 275, 281, 285, 286
Lotman, Mihhail 181, 182, 193
Lovelock, James E. 186, 187, 193
Lubischev, Alexandr 227, 268
Luisi, P. Luigi 111
Lück, Jacqueline 9, 17, 83, 86
Lyubischew, Alexander 227, 268

Maasik, Sonia 123, 128
Machado, Irene 184, 193
Magnus, Riin 269, 286
Malathrakis, Nikolaos E. 166
Mancuso, Stefano 85
Mandelker, Amy 181, 193

Manning, Peter K. 123, 128
Maran, Timo xii, 14, 19, 81, 88, 167, 168, 181, 175, 177, 286, 288
Marasov, Anatoli N. 284
Markellou, Emilia 166
Markoš, Anton xii, 7, 11–14, 19, 125, 128, 168, 170, 177, 184, 193, 235, 236, 241, 249, 250, 254, 255, 261, 287
Martinelli, Dario 14, 19, 80, 83, 88, 168, 177
Mathews, Cristopher 57, 65
Maturana, Humberto 114, 128
Mayr, Ernst 93, 110
McCaskill, John S. 109
McKee, Gerard T. 128
McLuhan, Marshall 188, 193
Melazzo, Lucio 20
Merrell, Floyd 14, 19, 119, 125, 127, 128, 190–193, 208, 261, 285
Meyen, Sergey 268
Meyer-Abich, Adolf 12, 19
Michison, N. Avrion 18
Middelburg, Jack J. 166
Millikan, Ruth G. 13, 19
Molin, Søren 165
Montesanto, Anna 87
Mooij, Wolf M. 166
Morri, Davide 87
Morris, Charles 171, 172, 177
Morris, Desmond 265, 286
Mounin, Georges 14, 19

Naumov, Rem V. 284
Nave, Ophir 197, 209
Neubauer, Zdenek 261
Neuman, Yair xii, 19, 20, 82, 88, 195–197, 202, 203, 205, 209, 245, 262, 288
Neumann, John von 228, 232
Newton, Isaac 7, 65
Nielsen, Michael A. 204, 209
Nielsen, Søren Nors 80, 88
Niklas, Ursula 13, 20
Nimchuk, Zachary 155, 166
Nöth, Winfried 11, 12, 20, 96, 107, 110, 184, 193

O'Hea, A. J. 161, 165
Ohgushi, Takayuki 146, 166
Ortega y Gasset, José 207
Osborn, Henry F. 140
Owen, Richard 8, 121

Packard, Norman H. 109, 111
Pain, Stephen P. 13, 20, 80, 83, 89
Palmer, Daniel K. 77, 89
Parniske, Martin 152, 166
Pasteur, Georges 175, 177
Pasteur, Louis 186
Paterson, Hugh E. H. 76, 89, 227, 274, 286
Pattee, Howard H. xii, 37, 40, 69, 89, 106, 110, 119, 128, 213–217, 219–226, 228–233, 246, 262, 288
Paul, Nigel D. 147, 156, 157, 166
Pearson, Karl 226, 228, 233
Peat, F. David 262
Peirce, Charles S. 13, 15, 17, 33, 34, 37, 40, 72, 89–91, 94–96, 98, 110, 111, 119, 129, 180, 190, 193, 200, 223, 238, 253, 258, 259, 262, 275–277, 281, 286
Penna, Maria P. 87
Pessa, Eliano 87
Petitot, Jean 125, 128
Petrilli, Susan 80, 88, 89, 115, 128, 129, 190, 191, 193
Piaget, Jean 90
Pjatigorski, Aleksandr M. 192
Plato 203
Plenio, Martin B. 204, 209
Polak, Michal 140, 141
Polanyi, Michael 69, 89, 115, 129, 195, 198, 199, 202, 209, 220, 233
Ponzio, Augusto 80, 89, 190, 191, 193
Portmann, Adolf 129, 170
Posner, Roland 15, 19, 20, 88, 90
Prigogine, Ilya 184, 193
Prodi, Giorgio 270, 286
Putten, Wim H. Van der 147, 148, 151, 152, 166

Queiroz, João 18, 80, 82, 86, 89, 109, 177

Raff, Martin 208
Ransdell, Martin 39, 85, 261, 284
Rashevsky, Nicolas 9, 10
Rasmussen, Steen 105, 107, 109, 111
Rauch, Irmengard 192, 193
Ray, Thomas S. 109
Redi, Francesco 180
Renoue, Marie 168, 177
Reynolds, Andrew 98, 111
Ribeiro, Sidarta 80, 89
Rice, Scott A. 150, 166
Richards, Stephen J. 163, 164
Roberts, Keith 208
Roepstorff, Andreas 82, 89
Rohmer, Laurence 166
Rolff, Jens 177
Roncadori, Roland W. 151, 166
Rosen, Robert 10, 20, 73, 89, 101, 111, 119, 123, 129, 227, 229, 233, 240, 246, 256, 262, 275
Rosenthal, Sandra B. 119, 129
Rothschild, Friedrich S. 2, 20, 25, 40, 72, 89, 123, 129, 227, 233, 270, 286
Russell, Bertrand 52, 53, 65
Ruyer, Raymond 249, 262

Saint-Hilaire, Etienne G. 8
Salmond, George P. C. 149, 166
Salthe, Stanley 111, 192, 283, 286
Sarkar, Sahotra 84, 87, 93, 110
Saussure, Ferdinand de 3, 6, 17, 277, 281
Saussure, Henri 6
Sbrocchi, Leonard G. 19, 128, 287
Schipani, Ileana 87
Schittko, Ursula 164
Schmid-Tannwald, Ingolf 82, 87
Schogt, Henry G. 5, 20
Schrödinger, Erwin 226
Schult, Joachim 14, 20, 89
Scozzafava, Silvia 87
Searle, John 75, 89, 93, 111
Sebeok, Thomas A. x, 1–3, 14, 16, 19, 20, 25, 26, 39, 41, 67, 69, 70, 80, 85, 88, 89, 90, 128, 129, 134–136, 141, 169, 170, 176, 177, 179, 189, 193, 196, 209,

294 Name Index

227, 232, 235, 236, 238–240, 259, 260–262, 269, 270, 275, 278, 281, 282, 284–286
Seddon, Barrie 150, 159, 166
Seibt, Johanna 110
Sen, Aparna 125, 128
Sercarz, Eli E. 18, 81, 90, 270, 286
Sériot, Patrick 9, 20
Sermonti, Giuseppe 7, 21, 268, 286
Shank, Gary 125, 129
Shapiro, Michael 89
Sharkey, Amanda J. C. 82, 90
Sharkey, Noel E. 82, 90, 107, 111
Sharov, Alexei A. 170, 171, 177, 227, 268
Sheets-Johnstone, Maxine 93, 104, 111
Sherman, Jeremy 75, 86
Sherratt, Thomas N. 175, 178
Short, Thomas L. 75, 90, 119, 129, 276, 277, 282, 286
Shrobe, Howard 127
Sibatani, Atuhiro 7, 18, 21, 268, 285, 286
Smith, Barry 125, 128
Smith, Jonathan Z. 11, 21
Solé, Ricard 9, 21
Solomon 203
Solomon, Jack 123, 128
Somorjai, Ray L. 246, 262
Spencer, Hamish G. 76, 88
Star, Susan Leigh 97, 111
Staskawicz, Brian 152, 166
Stengers, Isabelle 184, 193
Stent, Gunther 229, 233
Stepanov, Juri S. 14, 20, 21, 227, 233
Stewart, Gordon S. A. B. 165, 166
Stjernfelt, Frederik xii, 14, 18, 21, 25, 26, 31, 40, 41, 65, 79, 85, 87, 90, 98, 109, 111, 120, 129, 177, 235, 236, 239, 242, 243, 248, 251, 253, 257–259, 262, 263, 271, 284, 287, 288
Suess, Eduard 185
Suzuki, Hideaki 86, 127, 287
Swift, Simon 149, 166
Syernberg, Claus 165
Szostak, Jack 105, 111

Tada, Tomio 18, 90, 286

Takashi, Gojobori 164
Tamir, Boaz 202, 209
Taylor, Jane E. 166
Teilhard de Chardin, Pierre 185, 188, 193
Thom, René 7, 9, 21
Thompson, D'Arcy W. 8, 10, 21
Tiivel, Toomas 127, 285
Tiles, James E. 128
Tinbergen, Nikolaas 134
Toporov, Vladimir N. 192
Torop, Peeter 75, 78, 88, 124, 128, 184, 193, 269, 286
Trabant, Jürgen 41, 89
Turovski, Aleksei xii, 133, 288
Turovski, Markus 135

Uexküll, Jakob von 2, 9, 12, 17, 18, 21, 32, 33, 38, 40, 41, 73, 79, 81, 86, 90, 94, 110, 111, 119, 124, 129, 133–135, 167, 172–174, 176, 178, 180, 193, 196, 209, 224, 227, 240, 262, 268–270, 275, 276, 281, 285, 286
Uexküll, Thure von x, 39, 79, 85, 90, 98, 104, 111, 124, 129, 227, 261, 269, 270, 284
Ulanowicz, Robert E. 52, 65
Umiker-Sebeok, Jean 14, 19, 20, 25, 41, 128, 129, 177, 232, 236, 261, 262, 270, 285, 286
Uspenski, Boris A. 181, 192

Vane-Wright, Richard I. 168, 178
Varela, Francisco J. 128
Vehkavaara, Tommi 123, 129
Vernadsky, Vladimir 179, 181, 185–189, 192–194
Vet, Louise E. M. 166
Vetrov, Anatoli A. 14, 21
Vijver, Gertrudis Van de 94, 110, 111, 192
Villa, Alessandro E. P. 82, 90
Vitelli, Vincenzo 204, 209
Volkmann, Dieter 85
Vos, Matthijs 146, 166
Vrba, Elizabeth S. 45, 65

Wäckers, Felix L. 166
Waddington, Conrad H. 21, 52, 65, 128, 140, 227, 268, 286
Walter, Peter 208
Watanabe, Hidemi 164
Weber, Andreas 14, 21
Weber, Bruce H. 7, 17, 87, 261, 285
Webster, Gerry 7, 9, 18, 21, 285
West, Helen M. 151, 166
Wheeler, Wendy 14, 21
Whipps, John M. 151, 166
White, Harry C. 140, 141
Wicken, Jeffrey S. 101, 111
Wiener, Norbert 256, 262

Williams, Brooke 129
Williams, Paul 165, 166
Wilson, Dennis 151, 165
Wilson, Edward O. 264, 286
Winans, Stephen C. 166
Winson, Michael K. 165
Witzany, Günther 14, 17, 21, 86–88, 127, 128, 287, 288

Yates, F. Eugene 190, 194, 278, 286
Young, Evan P. 177, 286

Zarenkov, Nikolai A. 10, 21
Ziemke, Tom 82, 90, 107, 109, 111

Subject Index

abduction 33, 242
abiotic 38, 69, 145, 146, 151, 156, 157, 184
aboutness 38
absence 29, 30, 35, 75, 274, 276
action sign 32, 33; see also: *Wirkzeichen*
adaptation 12, 28, 33, 43, 55, 61, 64, 174, 265, 266, 275
adaptive behavior 25, 135
adaptive landscape 51
Aeromonas hydrophilia 163
agency 36, 45, 48, 50–53, 55, 61, 64, 73, 247, 282, 283
altruistic behavior 272
ambivalent sign 167, 169, 171, 172
analogue code 60, 157, 206, 271
animal communication 2–4, 14, 25, 35, 80, 83
animal-human communication 78, 133, 136
antagonist 147, 159, 162
anthropic 104, 105
anthropocentrism 257, 258
anthropology 8, 38, 99, 100, 136, 207
anthropomorphism 31, 134, 248
anthroposemiosis 104, 106, 108, 253, 257
Aphidius ervi 62
Aristotelian 79, 103, 237
artefact 223, 224
artificial life 91, 92, 94, 96, 99, 105–109
asemiotic 28, 43
asymmetric evolution 140
asymmetry 183
ATP 46

autocatalysis 28
autocell 27, 32
autocommunication 183, 184
autonomous agent 32, 102
autophytic 145
autopoiesis 97, 99, 102, 107, 203, 207
autozooic 145
awareness 38, 84, 219

bacteria 148–152, 161
bacterial communication 148–150
Baldwin effect 140
Bedeutungslehre 172, 173, 240; see also: *Theory of Meaning*
bees 45, 81, 174, 175, 272
behavioural ecology 4, 73, 74, 134
behaviourism 135, 171
Beobachtung 124
binarism 15, 183
binary distinction see next
binary oppositions 5, 11, 15, 182
bio-logic 13, 33
biobody 103–105, 108
biocomplexity 84
biodiversity 80, 84, 143, 144, 160, 226, 284
biofunctionality 102, 108
biohermeneutics 12, 125, 250
biolinguistics 13, 14
biological hierarchy 147, 149, 152, 157, 160–162, 195, 198–200, 203–205
biological information 147, 148, 238
biological structuralism 4
biophysics 68, 72, 114, 115, 123
biorhetorics 13

297

biosemantics 13
biosemiosis 73, 76–79, 96, 103, 105, 108
biosphere 16, 70, 115, 144, 179, 181, 182, 184–191
biotic 38, 69, 146, 156, 157
biparental reproduction 6
birds 62, 63, 74, 139, 173, 223, 267, 268, 272
body 9, 11, 47, 76, 77, 96, 99–106, 123, 170, 172, 173, 175, 186, 204, 205, 222, 238, 241, 242, 244, 249, 254, 277, 279–281; see also: biobody, embodiment
boundary (of the semiosphere) 182, 183
boundary condition 34, 69, 77, 195, 198, 199, 203, 207
brained animal 49, 62
Brobdingnagian 16
Brownian motion 98
Bufo boreas 163
Bufo bufo 174, 175

captivity 133, 134, 136
Cartesian 258
categorization 16, 37, 75, 78, 85, 226, 274, 283
causality 2, 32, 33, 59–62, 64, 69, 93, 94, 146, 148, 157, 184
cellomics 84
cellular 15, 25, 30, 59, 60, 83, 219, 246, 266,
cellularity 28
Central Dogma 237, 265
chemical information 146, 153, 155
chromosome 56–58
code 5, 6, 16, 28, 31, 64, 75, 81, 82, 100–103, 115, 117, 149, 201, 206, 207, 216, 218, 222, 242, 243, 274
code duality 181, 206, 207, 271
code-based 115
coding 3, 60, 170, 206
cognitive robotics 82
cognitive science 53, 91, 246
Coleoptera 140
collective agency 48

communication 2–4, 6, 11, 16, 25, 37–39, 59, 61, 70, 74, 75, 77–79, 81, 84, 92, 100, 104, 120, 123, 124, 140, 146–148, 150, 152–157, 159, 160, 168–171, 173, 174, 176, 180, 184, 190, 195, 200, 201, 206, 219, 221, 222, 226, 235, 242, 243, 268, 274, 280, 281
communicative 6, 39, 74, 75, 78, 117, 120, 126, 135, 136, 167, 168, 172, 272, 274
community 153, 242, 250, 259
complementarity 121, 222, 274, 275
complexity 9, 10, 95, 107, 122, 143, 144, 152, 153, 157, 160, 162, 196,
conditional causality 33
consciousness 38, 82, 93, 182, 188, 248, 253, 259, 273, 283,
consensual domain 114
constitutive absence 29, 35
constraint 8, 34, 213, 217–221, 225, 230
consumer 7
contextuality 1, 29
creodes 140
cross talk 28, 43, 55, 81, 146, 150
cue 28, 31, 38, 55, 59, 60, 62, 64, 151, 157, 158, 173
cultural selection 216, 230
cybernetic 186
cyborg studies 91

Darwinism 8
degeneracy 216
Developmental Systems Theory 78
diachronic 5
dialogue 114, 124, 182,
difference 16, 72, 74, 75, 79, 115, 119, 154, 200, 202–204,
differentiation 16, 27, 81, 84, 120,
digital code 60, 206, 271
digitality 59
directedness 44, 45, 101
disembodiment 106
DNA 51, 55, 56, 60, 69, 102, 153, 164, 195, 197–199, 201, 205–207, 229, 237, 254, 265, 271, 282
Drosophila 107, 140

dual opposition 119
dualism 2, 119, 258
dyadic 12, 94, 154, 157, 281

ecological complexity 143, 152
ecological niche 7, 38, 44, 48, 62, 95, 102, 107, 145, 146, 152, 271
ecological turn 252
ecology 6, 143, 145–148, 153, 242, 268
ecosystem 63, 80, 81, 143, 144, 147, 150–152, 154, 160, 161, 164
embodiment 38, 95, 96, 98–100, 102–107, 201
emergence, emergency 64, 97, 251, 255
emergentism 92
endophytes 153
entelechy 9
entropy 102
epigenetic 38, 52, 60, 84, 147, 152, 218, 221, 269
Escherichia coli 103
ethology 4, 62, 75, 78, 95, 133–135, 145–147, 242, 265, 267, 280
Evo-Devo 78, 100
evolution 29, 32, 34, 37, 39, 43, 46, 49, 52, 55–59, 61, 62, 64, 78, 79, 85, 93, 98, 103, 140, 143, 148, 154, 156, 164, 185, 187, 188, 213, 214, 219, 222, 228, 241, 247–250, 253–255, 266, 271–275, 282, 283
evolutionary biology 51, 265
evolutionary theory 78, 247, 265, 273; see also: theory of evolution
evolutionism 7
exaptation 45
extended mind 77

fidelity 28, 43, 55
Firstness 98
fish 58, 140, 173, 280
fitness 48, 63, 74, 98, 153, 156, 157, 168, 272, 279
Forficula auricularia 140
form 7–10, 34, 36, 37, 138, 199
form generation 36, 37
formalism 8

formative power 31
function 6, 9, 10, 25, 27–32, 34–36, 39, 43–49, 53, 55, 59, 60, 64, 74–76, 78, 102, 137, 154, 161, 167, 173, 214, 216, 231
functional circle/cycle 33, 73, 76, 77, 95, 224, 256
functionalism 8, 12
functionality 32, 36, 44, 59, 61
fungi 103, 150–152, 159, 163

Gaia hypothesis 186
gene 52, 57–60, 155, 158, 197, 207, 218, 219, 231, 265, 272
gene families 56
gene frequencies 49, 51
general biology 1, 27, 113, 268
general linguistics 1, 268
general semiotics ix, 3, 4, 33, 34, 252
generating relation 9
generative biology 9
genetic fixation 55, 61
genetic information 82, 95, 153, 197, 237, 265–267
genetic inheritance 37, 38
genetic memory 214, 231, 275
genetic replication 197
genome 55, 58–60, 82, 155, 206, 207, 219, 249
genome space 160
genomic memory 266
genomics 84, 231
genotype 59, 95, 266

habit-taking 12, 98
habitus formation 100
hemoglobin 28–30, 55–58
herbivore 143, 144, 149, 156, 158,
heritability 220, 264
Hirundo rustica 140
home range 135
homeostasis 99, 187
Homo sapiens 2, 140
homotypy 175
homunculus 43, 52–55, 64
host-symbiont relation 149

human intellect 25
human-animal communication 78
hybridization 105, 108
hybridize 97, 105
Hymenoptera 175

iconic 73, 79, 95, 102, 169, 170, 176, 181, 275
iconicity 169, 176,
idiographic method 126
immune system 60, 81, 82, 155, 163, 244, 270, 283
indexical 73, 79, 95, 102, 275, 281
individual distance 135
info-chemicals 143, 148
information 28, 31, 38, 43, 55, 64, 82, 91–96, 147, 148, 153–158, 164, 180, 182, 183, 195, 197–199, 202–206, 214, 215, 220–222, 229, 237, 238, 242–244, 246, 247, 250
information flow 146, 265, 266
informational ix, 16, 69, 143, 145, 147–149, 160, 203, 205–207, 238, 246, 266
inheritance 8, 37, 38, 221
insects 81, 140, 144, 151, 153, 162, 175, 176, 282
instinct 62, 134, 135
instruction 59, 78, 257
intentionality 1, 12, 75, 83, 103, 247, 283
interpretance 223
interpretant 33, 34, 36, 94, 96, 119, 156, 158, 171, 244, 248, 259
interpretation 1–3, 5, 13, 36, 37, 39, 59, 69, 71, 94–96, 115, 117–122, 167, 170, 172, 206, 244, 250
interpretative agency 48
interpreter 35, 39, 82, 96, 135, 171, 172, 213, 223
interpretive capacity 35
interpretive challenge 38
intracellular 157, 282
inverse sign 170, 171
invertebrates 83, 151
irreversible systems 64
irritability 100, 103, 283

language capacity 3
law-equivalence 216, 219, 220
lawfully indeterminate 215
learning 12, 46, 48, 49, 55, 61–63, 74, 75, 146, 156, 220, 221
Lebenswelt 97, 100
legisign 95
life process 15, 27, 67, 77, 84, 91, 97, 106, 113, 114, 119, 179, 181, 185, 187, 190, 191, 219, 237, 274, 276
lineage 48, 57, 58, 103, 107, 249, 250, 266, 282
linguistics 1, 3, 5, 9, 11, 14, 121, 268,
living system 1, 2, 11, 14–16, 28, 36, 59, 60, 64, 69, 72, 73, 75, 82, 92, 95, 102, 113–115, 119, 120, 125, 126, 195, 196, 198, 202, 203, 205, 216, 219, 226, 282
Lophius piscatorius 173

macro-herbivores 151
mammals 29, 58, 62, 63, 139, 150, 155, 272
Manduca sexta 158
matter-mind 2, 103, 222
meaning ix, 95, 96, 116, 154, 167, 183–185, 195, 196, 201, 202, 230, 258, 267, 270, 274, 276, 278–280
meaning carrier 172, 173
meaning triggers 196
meaning-making ix, 80, 196, 202
mediation 1, 94, 197, 201, 273
memory 38, 62, 75, 102, 140, 156, 182, 213–215, 219–223, 225, 242, 249, 274, 283; see also: genetic memory
Merkzeichen 119
Mertensian mimicry 138
message 47, 69, 156, 181–183, 196, 225, 243, 266
messenger 28, 43, 55, 81, 158
metabolism 16, 29–31, 46, 99, 102, 158, 225
micro-herbivores 151
mimesis 123
mimic 167–169, 171, 172, 174–176
mimicry 81, 138, 167–176

mind 2, 11, 39, 91–93, 96, 99, 103, 179, 187–189, 200, 222, 257–259, 275, 276, 278, 279
mind and matter see: matter-mind
mind-reading 74, 75
mindedness 84, 257
modelling relation 229
modelling 81, 120, 121, 181
Modern Synthesis 68
molecular biology 28, 82, 83, 94, 146–148, 229–231, 242, 245, 246, 266
molecular chemistry 28
monist 113, 227
Mormoniella vitripennis 140
morphogenesis 84
morphogenetics 10
morphology 7, 10, 11
Müllerian mimicry 174
multicellular 29, 48, 100, 103, 104, 221
multicellularity 99, 283
multitrophic interaction 143, 144, 147, 148, 151, 152
multitrophic systems 143–164
mutation 9, 14, 19, 135
mycorrhiza 150, 151, 153, 159
myoglobin 57

natural selection 8, 32, 43–45, 48, 51–53, 55, 56, 61, 74, 92, 214, 216, 219, 220, 230, 266, 272, 273, 275
need for impression 137, 138
neo-Darwinist/neodarwinian 28, 39, 51–53, 227, 254, 255
nervous system 103, 104, 218, 222, 244, 283
neurobiology 15, 95
neurosemiotic 82
Nicotiana attenuata 153, 158
nomothetic 126
non-deterministic 64
non-human 2, 53, 64, 105, 123, 252, 253, 260
non-mechanicism 92
non-semiosic 3, 27, 278, 281
non-semiotic 77, 79, 85, 182, 198, 199, 275

non-verbal signs 3, 198
non-vitalism 92
nonfunctional 55, 58
nonliving 69
noosphere 179, 185, 187–189
normativity 36, 37, 97
nucleic acid 51, 215

observation 124
ontogeny 34, 53, 60, 254
ontology 6, 97, 98, 100, 106, 113, 267
open-ended evolution 213, 214
Ophrys insectifera 174
organicist 84, 91–93, 96, 109
organization 4, 10, 27, 32, 36, 78, 96, 100, 180, 186, 187
organizing self 68
Osaka Group 7, 268
Osborn-Baldwin effect 140
oxygen 29, 30, 46, 56–58, 163

pace 37
parasitoid 143, 144, 149
participation 124
Passilidae 140
pathogens 143, 144, 147–152, 158–161, 163, 164
Peircean 26, 33, 36, 73, 82, 110, 115, 190, 242, 252, 253, 259, 271, 273
perception sign 32, 33; see also *Merkzeichen*
perception-action cycle 104; see also: functional cycle
phenomenological 71, 95, 96, 125
phoneme 6
photosynthesis 283
physical laws 1, 6, 115, 120, 213–216, 218–220, 222, 230, 231, 277
physicalist science 14, 67, 68, 122, 123, 125, 238, 247
plant signaling 62
plants 62, 81, 83, 103, 109, 143–160, 162, 172, 174, 190, 282
pluralism 119, 122
pluralist 113, 119, 277
plurireal 126

Subject Index

polypeptide 10
population 44, 48, 51, 107, 108, 137, 140, 146, 159, 163, 221, 241, 266, 271, 274
Portmannian 170
posthuman 105
preadaptation 45
prebiotic 27, 55
predator-prey relation 62
producer 7
prokaryotes 148
proteomics 84, 231
proto-argument 33
proto-life 27
protozoa 57, 103
pseudogene 55–57
psychology 7, 95

qualisign 95
qualitative difference 74, 75, 79, 119, 121, 224, 274
qualitative diversity 16, 72, 80, 274
quorum sensing 143, 148–150, 159

receptors 47, 60, 124, 282
recognition 6, 15, 16, 34, 75, 76, 102, 116, 121, 137, 139, 146, 153–158, 167, 168, 171, 176, 221–227, 242, 271, 274
reductionist 64, 68, 70, 83, 94, 123, 146, 190, 238, 244, 245
redundancy 81, 216
reflex 51, 62
reflexivity 99, 104
relation 1, 62, 145, 154, 171, 222–224, 229, 248
relational biology 9, 229
relationality 5, 6, 10, 29, 30, 113, 116, 121
reproduction 6, 8, 45, 46, 49, 99, 108, 115, 140, 221, 226, 241, 266, 272, 283
response 48, 61, 78, 168, 174, 195, 196, 201, 202, 220, 272
Rhodeus 173
RNA 82, 102, 197, 201, 204, 206
robots 106–108

salamanders 265, 266
Salmonella 163
Saussurean 5, 8, 277
scaffolding 34, 55, 58–62, 253, 254
Sebeok's thesis 69
Secondness 273
self 16, 32, 81, 219, 257
self-maintenance 30, 32, 24
self-organization 31, 68, 69, 92, 100–102, 184, 185, 218, 251
self-referentiality 35, 282
self-regulation 184, 185, 187
self-replication 28, 228
self-reproduction 32, 93, 108
semethic interaction 189
semiobiosphere 190, 191
semiogenic scaffolding 59
semiology 1, 3, 5, 12, 277
semiosic 3, 11, 13, 15, 27, 75, 77, 81, 84, 133, 179, 189, 273–275, 281, 282, 284
semiosic matrix 82
semiosis 1, 2, 5, 11, 15, 16, 25, 27, 28, 32, 33, 35–38, 43, 55, 64, 68–70, 73, 75–80, 83, 94, 96, 103–108, 115, 117, 118, 120, 125, 126, 137, 168, 171, 179, 180, 182, 189–191, 221–224, 226, 230–232, 247–248, 273–275, 278, 281–283; see also: anthroposemiosis, biosemiosis
semiosphere 70, 135–137, 140, 163, 179–184, 188–191, 226, 269, 275, 281
semiotic balance 80, 284
semiotic border 117
semiotic control 60, 152, 213, 223–225
semiotic emergence 272, 278
semiotic freedom 63, 213, 216, 271, 279, 282
semiotic models 12, 81, 121, 191
semiotic mutualism 63
semiotic niche 38, 39, 62, 145, 150, 163, 164, 271
semiotic process 2, 15, 28, 33, 35, 36, 38, 48, 80, 81, 118, 124, 151, 152, 162, 164, 183, 189, 243, 248, 254
semiotic subject 282
semiotic threshold 25, 79, 224, 275, 281–283

semiotic trait 140
semiotics of culture 123, 281
sensing 73, 81, 93, 143, 145, 146, 148, 149, 152, 156–159, 161–163, 190
sexual reproduction 45, 283
sign action 5, 25, 33, 94–96, 99, 223, 246, 259; see also: semiosis
sign family 171, 172
sign process 2, 4, 12, 15, 25, 27, 30, 34, 61, 67, 68, 72, 73, 78, 82, 83, 91, 95, 96, 98, 102, 123, 126, 180, 181, 191, 216, 222, 250; see also: semiosis, sign action
sign production 1, 35, 154
sign relation 11, 12, 16, 70, 72, 73, 77, 78, 80, 171, 191, 235, 248, 257
sign system 1–4, 15, 25, 34, 78, 79, 97, 101, 125, 180
sign vehicle 36, 37, 60, 94, 96, 102, 171, 172
sign-mediated activity 196, 197, 202, 203, 206
signal 28, 30, 31, 43, 45, 55, 60, 62–64, 81, 82, 93, 103, 135, 137, 143, 146, 148–150, 152, 154, 155, 157, 158, 162, 163, 168, 170, 190, 246
signal transduction 60, 81, 82, 103, 143, 148, 158, 282
signal-receiver 167–172, 174, 176
signification ix, 1, 5, 6, 25, 32, 68, 69, 105, 148, 190, 195, 196, 200, 201
significatum 172
signosphere 189
signum-structure 135
similarity 6, 168, 171–175, 242, 274
sinsign 95
situated interpreter 82
social sign system 25
sociobiology 264, 265, 267, 272
sociology 7, 99, 100
speciation 6, 16, 135, 266, 274
species 6, 39, 50, 52, 75, 76, 120, 121, 140, 168, 221, 226, 227, 266, 271, 274, 282
species coexistence 268
species-specific 70, 72, 135, 254

stability 16, 38, 152, 161, 184, 185, 205–207, 214, 226, 253, 284
stereochemical relationship 3
stimulus 5, 220
stimulus-reaction 135
striving 49, 50, 53, 55, 283
structuralism 1, 4–9, 11, 12, 15, 252
subject 48, 95, 102, 138, 140, 156, 226, 247, 266, 267, 271, 282,
subjective category 160–162, 164
subjective experience 70, 71, 93, 140, 246
subjectivity 84, 93, 126, 140, 247, 271
succession 6, 48
symbiosis 62, 63, 149, 150, 161, 164
symbiotic 62, 150
symbol 15, 37, 95, 222, 223, 230, 231, 243, 279
symbol-matter problem 222, 223, 231
symbolic 3, 17, 33, 73, 79, 102, 213, 218, 221, 228, 230, 231, 275, 281
symbolic threshold 79, 275, 281
symbolization 279
synchronic 5
Syrphidae 175

taxonomy 6, 37, 82, 169, 173, 174
Teilnahme 124
teleo-functionality 36
teleological 15, 27, 31, 32, 44, 49, 52, 64
teleology 31, 44, 92, 93, 95, 101, 105
teleonomy 44
telos 27, 31, 32
theoretical biology 1, 25, 26, 31, 92, 113, 196, 227, 231, 238, 267, 268
theory of complexity 122
theory of evolution 16, 39, 59–63, 139
Theory of Meaning 167, 240; see also: *Bedeutungslehre*
Thirdness 79, 98, 273
threshold zone 25, 27, 28, 55
transcription 55, 60, 82, 148, 153, 158, 237
transcriptomics 84, 231
transformation 8, 145, 197–199, 202, 204

transgenerational subject 48
translation 3, 60, 78, 82, 115, 124, 125, 182–184, 198, 237, 275
transmutation 198, 199
triadic 11, 12, 94, 154, 156, 157, 243, 281
triadic process 94
triadic relation 11, 154, 157, 223, 229
triadic sign relation 77
Turing machine 205

Uexküllian 73, 167, 172–174, 176
Uexküllian question 67, 72
umwelt 2, 25, 38, 39, 70–75, 81, 82, 95, 99, 104, 109, 114–117, 124, 133–138, 140, 164, 167, 172–176, 220, 246, 247, 257, 269
unireal 126

variation 8, 92, 220
vegetative semiosis 83
vertebrates 82, 83, 151
vervet monkey 80
Vicia faba 62
virulence 150, 155, 160–164
viruses 151
vitalism 71, 93, 248, 249

wasps 174, 175
wholeness 1
Wirkzeichen 119; see also: action sign

zoology 100, 104
zoopsychological 134
zoosemiotics 2, 14, 168, 270, 281